Lecture Notes in Mechani

Lecture Notes in Mechanical Engineering (LNME) publishes the latest developments in Mechanical Engineering—quickly, informally and with high quality. Original research reported in proceedings and post-proceedings represents the core of LNME. Also considered for publication are monographs, contributed volumes and lecture notes of exceptionally high quality and interest. Volumes published in LNME embrace all aspects, subfields and new challenges of mechanical engineering. Topics in the series include:

- Engineering Design
- Machinery and Machine Elements
- Mechanical Structures and Stress Analysis
- Automotive Engineering
- Engine Technology
- Aerospace Technology and Astronautics
- Nanotechnology and Microengineering
- Control, Robotics, Mechatronics
- MEMS
- Theoretical and Applied Mechanics
- Dynamical Systems, Control
- Fluid Mechanics
- Engineering Thermodynamics, Heat and Mass Transfer
- Manufacturing
- Precision Engineering, Instrumentation, Measurement
- Materials Engineering
- Tribology and Surface Technology

More information about this series at http://www.springer.com/series/11236

Mokhtar Awang
Editor

The Advances in Joining Technology

Editor
Mokhtar Awang
Department of Mechanical Engineering
Universiti Teknologi PETRONAS
Seri Iskandar
Malaysia

ISSN 2195-4356 ISSN 2195-4364 (electronic)
Lecture Notes in Mechanical Engineering
ISBN 978-981-10-9040-0 ISBN 978-981-10-9041-7 (eBook)
https://doi.org/10.1007/978-981-10-9041-7

Library of Congress Control Number: 2018938358

Printed on acid-free paper

This Springer imprint is published by the registered company Springer Nature Singapore Pte Ltd.
part of Springer Nature
The registered company address is: 152 Beach Road, #21-01/04 Gateway East, Singapore 189721, Singapore

Contents

Thermal Modelling of Friction Stir Welding (FSW) Using Calculated Young's Modulus Values 1
Bahman Meyghani, M. Awang, S. Emamian and Mohd Khalid B. Mohd Nor

The Effect of Pin Profiles and Process Parameters on Temperature and Tensile Strength in Friction Stir Welding of AL6061 Alloy 15
S. Emamian, M. Awang, F. Yusof, Patthi Hussain, Bahman Meyghani and Adeel Zafar

The Effect of Argon Shielding Gas Flow Rate on Welded 22MnB5 Boron Steel Using Low Power Fiber Laser 39
Khairul Ihsan Yaakob, Mahadzir Ishak and Siti Rabiatull Aisha Idris

Effect of Bevel Angle and Welding Current on T-Joint Using Gas Metal Arc Welding (GMAW) 49
Z. A. Zakaria, M. A. H. Mohd Jasri, Amirrudin Yaacob, K. N. M. Hasan and A. R. Othman

Laser Brazing Between Sapphire and Inconel 600 59
Shamini Janasekaran, Farazila Yusof and Mohd Hamdi Abdul Shukor

A Review on Underwater Friction Stir Welding (UFSW) 71
Dhanis Paramaguru, Srinivasa Rao Pedapati and M. Awang

Three Response Optimization of Spot-Welded Joint Using Taguchi Design and Response Surface Methodology Techniques 85
F. A. Ghazali, Z. Salleh, Yupiter H. P. Manurung, Y. M. Taib, Koay Mei Hyie, M. A. Ahamat and S. H. Ahmad Hamidi

A Review on Mechanical Properties of SnAgCu/Cu Joint Using Laser Soldering 97
Nabila Tamar Jaya, Siti Rabiatull Aisha Idris and Mahadzir Ishak

Thermal Modelling of Friction Stir Welding (FSW) Using Calculated Young's Modulus Values

Bahman Meyghani, M. Awang, S. Emamian and Mohd Khalid B. Mohd Nor

Abstract The temperature fluctuations present in Friction Stir Welding (FSW), require, detailed thermal analysis of the process. To achieve highly accurate results for the analysis, reliable material data should be obtained. Nevertheless, the material data that are presented in the literature are usually limited to lower strain rate regimes and lower temperatures. Thus, calculating the temperature dependent material properties helps improve the accuracy of the stimulated model. To achieve a reliable material data in the higher range of temperatures, this paper presents a mathematical formulation for calculating temperature dependent Young's modulus values. MATLAB® and ABAQUS® software are employed for solving the governing equations and modelling the process, respectively. To compare the results and find the error percentage, the calculated and the documented (constant) values of Young's modulus are applied into two distinct finite element models. Finally, the developed model is validated by comparing the results obtained from experiments with published results.

Keywords Temperature fluctuations · Friction stir welding (FSW) · Mathematical formulation · Young's modulus

1 Introduction

FSW combines both mechanical and thermal phenomena. Basically, the temperature evaluation surrounding and inside the stirring zone (SZ) is dominated by severe plastic deformation and frictional force [1]. To clarify the point, plastic deformation produces mechanical energy and some parts of this mechanical energy is transformed

---

B. Meyghani · M. Awang (✉) · S. Emamian
Department of Mechanical Engineering, Faculty of Engineering, Universiti Teknologi PETRONAS, Bandar Seri Iskandar 32610, Perak Darul Ridzuan, Malaysia
e-mail: Mokhtar_awang@utp.edu.my

M. K. B. Mohd Nor
Friction and Forge Processes Group, Joining Technologies Group, TWI Ltd, Cambridge, UK

M. Awang (ed.), *The Advances in Joining Technology*, Lecture Notes in Mechanical Engineering, https://doi.org/10.1007/978-981-10-9041-7_1

into heat [2]. This issue causes permanent microstructural changes and also changes the phase of the material.

Many studies have been done to investigate the temperature behaviour during the FSW process by using experimental procedures [3–5], however, for detailed understanding and analyse the thermal behaviour, experimental methods are usually costly and time consuming. Finite Element Methods (FEMs) is recommended as a powerful tool and an effective numerical technique for solving partial differential equations (PDEs) in engineering, therefore FEMs are a cheaper and quicker approach for investigating the process. A schematic view of the FSW is described in Fig. 1.

As discussed earlier, the most significant factors for the generation of the heat are frictional force and plastic deformation [6, 7]. Therefore, to obtain more precise analysis of the process, it is crucial to precisely define the material plasticity behaviour. One of the most significant input parameters which is influencing the accuracy of the simulated model is Young's modulus. However, Young's modulus that were used in the literature, usually obtained from the experimental data or gained from the literature [8–10]. Additionally, the values are often restricted to lower strain rate regimes and lower temperatures. While, temperature in FSW can reach up to 60–80% of the base material melting temperature [11].

In recent years, the FSW process modelling has been researched in both academic and industrial organisations. A Lagrangian–Eulerian (adaptive arbitrary formulation) was established to compute the evolution of the temperature and the flow of the material throughout the FSW process [12]. In the model, Young's modulus was kept constant at 73 GPa and the Poisson's ratio was assumed to be 0.3. In this research, 3D Forge3® FE software with automatic remeshing was utilised to implement the formulation. The results indicated that the accuracy of the model in the tempera-

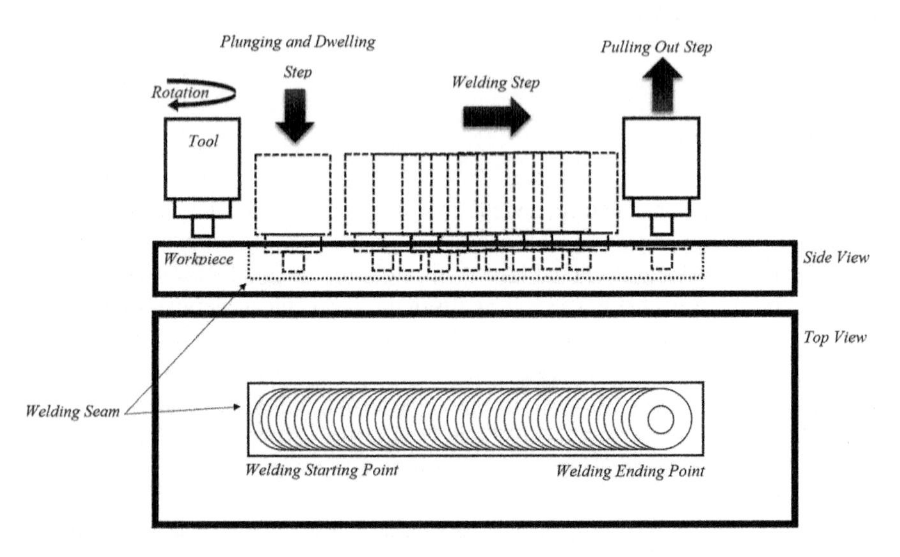

Fig. 1 The process schematic view and rotation direction

ture prediction is highly dependent to the use of constant or temperature dependent material properties.

The arbitrary Lagrangian–Eulerian (ALE) method was used to develop a 3D FE model in ABAQUS®/Explicit by utilising Coulomb law of friction and Johnson—Cook material law [8]. The study found that the generation of the heat during the process can be categorised into two various parts, the heat generated by the deformation of the material near the pin and the shoulder, and the frictional force. In addition, the numerical results of the paper indicated that the heat generated by the friction has more impact in comparison to the heat produced by the plastic deformation. The study also demonstrated that some portions of the heat is the result of the slip rate. Meanwhile, it was claimed that, the heat from the plastic deformation is linked to the material velocity.

To estimate the residual stresses and the peak temperature, a stationary shoulder was used in a study [9] in which a constant Young's modulus and Poisson's ratio of 71.7 and 0.33 GPa have been employed, respectively. It was observed that, through the thickness of the material, the process produced a more uniform and a narrower nugget zone and the heat affected zone. Furthermore, 'M' shaped residual stress distributions were obtained.

Meanwhile, a finite element (FE) analysis was employed for simulating the stress, the effective strain distribution and the temperature on the workpiece surface for different pin profiles [10]. The model used a constant elastic modulus of 68.9 GPa and a constant Poison's ratio of 0.3. The study examined the fracture surfaces at both microscopic and macroscopic levels to investigate the welded joints fracture behaviour. It was found that the crack initiated from the periphery of the joint, and the failure of the joints mainly occurred due to the thinning of the upper sheet.

A study presented [13] a solid approach by using a constant Poisson's ratio of 0.33 and Young's modulus of 70.4 GPa for thermomechanical simulation of FSW. Two numerical models (three dimensional models) were compared representing FSW procedures with a trigonal pin. One of the models was based on a solid formulation, while the other one was based on a fluid formulation. Authors used Norton–Hoff constitutive model with high sensitivity to the temperature, and the Arbitrary Lagrangian Eulerian (ALE) method. It was concluded that, the two mentioned formulations lead to the same results.

In a model [14] a constant poison ratio of 0.33 and Young's modulus of 68.9 GPa were applied to the model. The paper proposed a 3D coupled thermo-mechanical model, which was based on the Lagrangian implicit approach to examine the distribution of the strain and also to observe the thermal history of Aluminium alloy 2024 butt welding using the DEFORM-3D® software. It was observed that there is an asymmetric nature in the welding nugget zone. Furthermore, it was found that the top surface of the workpiece has the maximum temperature.

In the meantime, some studies adapted temperature dependent material properties [15–18]. The material characteristics were introduced [15] to model the process using the constant Poisson's ratio of 0.33, and the temperature dependent Young's modulus ranging from 68.9 GPa in 25 °C–31.72 GPa in 426.7 °C. Moreover, in the model the relationship between the tool moving speed, the heat distribution and the residual stress were investigated. The analysis of the process was classified into two stages; the first stage involved the studying of the workpiece thermal behaviour where the heat was generated because of the friction between the interfaces of the tool/workpiece. Meanwhile, in the next stage, the workpiece thermal behaviour (investigated from the first stage) was taken into account as an inlet heat which represents the elasto-plastic behaviour. In the latter step of the simulation, the tool was removed after the welding and the distributions of the residual stress was measured after completely cooling of the workpiece (when the clamp was disassembled). The results obtained showed that, the pattern of the distribution of the heat varied along the thickness is largely asymmetrical. Furthermore, it was observed that, as the welding transverse and rotational movements increased, the welding longitudinal residual stress also increased. It should be noted that, only the heat impact was considered for predicting the residual stress. Therefore, it was deemed as the main factor of the minor differences between the actual experiment and the simulation.

A study had developed a 3-dimensional localised FEM for predicting the probable results for the defects generation within the FSW [18]. The Lagrangian formulation had been used to model the tool, while the Eulerian formulation was used to model the workpiece. Besides that, the Coulomb law of friction was used for defining the interactions between the interfaces of the tool/workpiece. Moreover, the material inflow and outflow velocities were used to define the welding speeds. The study had considered the temperature dependent Young's modulus, ranging from 66.94 to 20.2 GPa when the temperature variation is 25–482 °C. The results investigated that by considering the adiabatic heating effect, the maximum estimated temperature of 583 °C was obtained, which was similar to the material solidus temperature (based on Johnson–Cook material model).

The temperature dependent Young's modulus was adopted to investigate the mechanical behaviour of the material during high temperature for welding of 6xxx aluminium alloy series [17]. Based on the result, the temperature had continuously changed during the welding; firstly, the temperature had risen, then it decreased (during the cooling down period). It should be noted that, the dwell-time at the maximum temperature was not considered in the model. The study also examined, the tensile tests during the heating and the cooling rate of the specimen. Moreover, a comparison between the calculated and the measured stress-strain curves was done to validate the accuracy of the thermomechanical database. This suggests that, this approach could be useful in predicting the alloys tensile behaviour at high temperatures. Lastly, the numerical and the experimental results for the temperature and the residual stresses were compared and a good agreement was found.

Meanwhile, Aziz et al. [16] employed Young's modulus as a temperature function in a study for developing thermomechanical modelling FSW of aluminium (AA2219)–copper alloy. Furthermore, they investigated the heat generation throughout the process. The model was developed by using ANSYS® APDL and the results were verified through conducting a comparison of the temperature profile between the experimental observations and the results obtained from the simulated model. Three different conditions (various welding speeds) in experimental and numerical models were considered and the verified FE model was used to analyse the influence of the welding parameters in the heat generation. It was observed that, the influence of the rotational velocity in the heat generation is higher than the transverse speed.

Although, as cited in previous paragraphs, the input materials parameters which have been used in the literature were usually derived from the experimental tests or the literature, while the experimental methods are generally costly and time-consuming. Moreover, it should be noted that the inaccuracy was always observed in the experimental measurements. Besides, the available material data in the literature was often restricted to lower temperatures, and lower plastic deformation. This is in contrast to what happens in reality during the FSW where there is a high range of temperature variations and large plastic deformation. Furthermore, the material behaviour for high temperature is very complicated and cannot be derived easily. Therefore, it is quite challenging to find high temperature material properties, and also for some materials, it might not be existing at all. Thus, researchers are still looking for the precise calculation of the Young modulus that can be obtained through straightforward methods.

Hence, this paper proposes a mathematical formulation that can be employed to calculate the temperature dependent values for the Young modulus. Moreover, in this paper, the calculated and the documented values of the Young modulus are employed in two different FE models in order to investigate the influence of the Young's modulus in the thermal behaviour of FSW. Finally, to validate the calculated values, the obtained results have been compared with both experimental observations and published papers.

2 Methodology

2.1 Mathematical Model

In this part, the mathematical formulations for deriving the temperature dependent Young's modulus is described. The shear stress (τ_y) was written based on the von Mises yield criterion, as follows,

$$\tau_y = \frac{\sigma_y}{\sqrt{3}} \tag{1}$$

where σ_y represents the yield stress,

In materials science, the modulus of shear or the modulus of rigidity is denoted by G, which refers to the ratio between the shear stress and the shear strain as follows,

$$G = \frac{\tau_{xy}}{\gamma_{xy}} \tag{2}$$

where the shear stress is $\tau_{xy} = \frac{F}{A}$ in which F represents the acting force and A represents the area wherever the force acts, and the shear strain is $\gamma_{xy} = \frac{\Delta_x}{l}$ in which Δ_x is the transverse displacement and l is the initial length, consequently,

$$\tau_{xy} = \frac{G\,\Delta_x}{l} \tag{3}$$

By incorporating Eq. (1) into Eq. (3), the yield stress value can be derived as,

$$\sigma_y = \frac{\sqrt{3}\,G\,\Delta_x}{l} \tag{4}$$

Furthermore, based on the conversion formulas, the shear modulus can be written as,

$$G = \frac{E}{2\,(1+\vartheta)} \tag{5}$$

where E represents the Young's modulus, or known as the modulus of elasticity and ϑ represents the Poisson's ratio. By incorporating Eq. (5) into Eq. (4), the yield stress value can be obtained as follows;

$$\sigma_y = \frac{\sqrt{3}\,E\,\Delta_x}{2\,(1+\vartheta)\,l} \tag{6}$$

Therefore, the Young's modulus can be calculated as follows,

$$E = \frac{(2l\,(1+\vartheta))\;\sigma_y}{\sqrt{3}\,\Delta_x} \tag{7}$$

where σ_y (yield stress) will be calculated using the Johnson-Cook material model,

$$\sigma_y = \left[A + B\,(\varepsilon_P)^n\right]\left[1 + C\left[\frac{\dot{\varepsilon}_P}{\dot{\varepsilon}_0}\right]\right]\left[1 - \left[\frac{T_{FSW} - T_{room}}{T_{melt} - T_{room}}\right]^m\right] \tag{8}$$

where ε_P, $\dot{\varepsilon}_P$ and $\dot{\varepsilon}_0$ represent the effective plastic strain, the effective plastic strain rate and normalizing strain rate respectively. Moreover, the room temperature (T_{room}) is 25 °C and the AA6061-T6 melting temperature (T_{melt}) is 582 °C. A represents the yield stress (546 MPa), B represents the strain factor (678 MPa), n represents the strain exponent (0.71), C represents the strain rate factor (0.024) and m represents the temperature exponent (1.56).

2.2 *Finite Element Modelling and Geometry*

The FE model and the refined mesh near the stir zone are illustrated in Fig. 2. The pin length was set to be 8 mm, while the diameter for the shoulder was 24 mm and the shoulder plunge depth into the workpiece was considered to be 0.15 mm. It should be noted that, a workpiece with a dimension of 200 mm (length) × 100 mm (width) × 10 mm (thickness) was drawn.

ABAQUS® FE software is used for simulating the process, because it is recommended as a suitable approach [19]. ALE formulation was used in the stirring zone to prevent extremely large mesh distortions, while the Eulerian formulation was used to fix the mesh of the other areas of the work piece. In addition, the three-dimensional coupled thermomechanical hexahedral element (C3D8RT) with 8 nodes were used, because in this kind of element nodes contains degrees of freedom and trilinear displacement. Moreover, these elements have an hourglass control and a uniform strain (the first order reduced integration). It needs to be mentioned that, the welding tool has been assumed as a rigid body in which thermal degrees of freedom was not considered. Furthermore, the temperature dependent values for the coefficient of friction in a variety of 0.207089–0.00058, for the variation of the temperature between the room temperature to the AA6061-T6 melting point obtained from the literature [20–22]. A total of 9 models are studied in which three welding rotational speeds (800, 1200 and 1600 RPM) and three different transverse speeds (40, 70 and 100 mm/min) are employed. Moreover, the tool material was steel H13 and AA6061-T6 was selected as the material for the welding plate. Tables 1 and 2 indicate all of the materials properties and the process parameters respectively.

Printed using Abaqus/CAE on: Wed Feb 15 14:31:01 Malay Peninsula Standard Time 2017

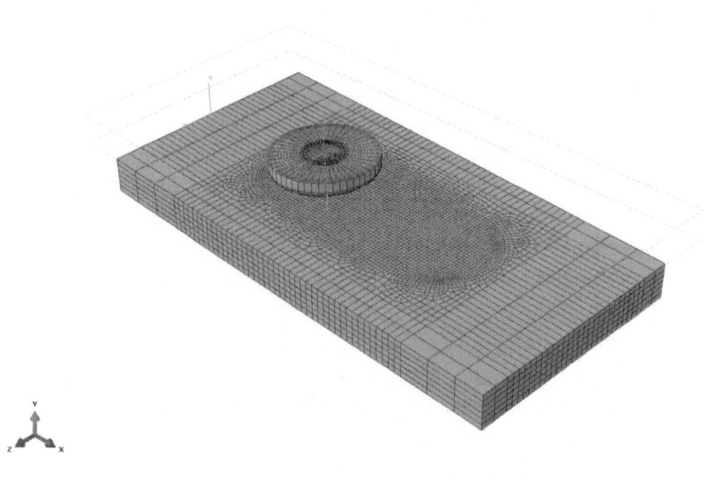

Fig. 2 The mesh and the boundary condition for the welding plate and the tool

Table 1 Temperature dependent material properties

Temp (°C)	Coefficient of thermal expansion (°C)	Specific heat capacity (J/Kg °C)	Density (kg/m^3)
37.8	2.345E−005	95	2685
93.3	2.461E−005	978	2685
148.9	2.567E−005	1004	2667
204.4	2.66998E−005	1028	2657
260	2.756E−005	1052	2657
315.6	2.853E−005	1078	2630
371.1	2.957E−005	1104	2630
426.7	3.071E−005	1133	2602

Table 2 Mechanical properties of AA6061-T6 (temperature dependent)

Temp (°C)	Thermal conductivity (W/mK)	Poisson's ratio
148.9	162	0.34
204.4	177	0.34
260	184	0.34
315.16	192	0.34
371.1	201	0.34
426.7	207	0.34
148.9	217	0.34
204.4	223	0.34

3 Results and Discussions

3.1 Mathematical Formulation Results

Basically, the initial point for the deformation of the material is highly depending on the material shear stress, while the material shear stress highly depends on the temperature. To clarify the point, the material yield stress point denotes the beginning stretch point of the material. This issue means that, the deformation of molecules or atoms inside the material will begin, if the shear stress of the material surpasses to the yield stress point of the material. On the other hand, as mentioned earlier, it should be also considered that the material final yield stress point highly depends on the process temperature.

Figure 3 shows the Young's modulus values calculated for the whole range of the temperatures during the process. As can be seen, the values are compared with published results [14–16]. The results show that, the values are very close to what has been reported in the literature, proving the validity of the Young's modulus calculated in the present work. This illustrations that the model can be used as a valid model for introducing the relationship between the temperature and the Young's modulus

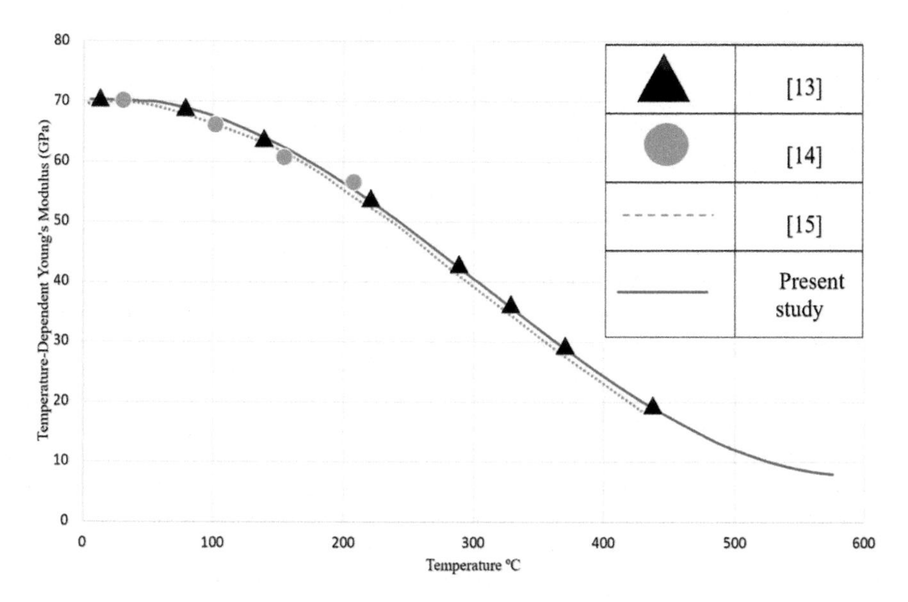

Fig. 3 A comparison between the calculated values of the Young's Modulus with the published results ([13–15])

for aluminum 6061-T6, in the temperature ranging from the room temperature to the material melting point (580 °C). From the result, it can be concluded that as the temperature increases, the Young's modulus values fell in the range of 70–0.8 GPa.

3.2 *Finite Element Results*

Two sets of the material properties were applied in the model for examining the effect of the temperature-dependent material properties of AA6061-T6 in the transient temperature. In the model one, constant Young's modulus value (71.7 GPa) was applied while in the second one temperature dependent values were applied. The results of the two numerical models were compared with the results obtained in experimental observations.

3.3 *Error Percentage Results*

Table 3 presents the comparison between the percentage of the error in two FE models with experimental measurements [20]. As shown, the error percentage in the second model is decreased significantly.

Table 3 The percentage of the error in different rotational (RPM) and transverse (mm/min) velocities

Welding parameter	Model type	Temperature °C	Absolute error %
800 RPM–40 Mm/min	Experiment	295.828	–
	Model 1	261.312	13.2
	Model 2	301.267	1.83
800 RPM–70 Mm/min	Experiment	285.366	–
	Model 1	268.107	6.4
	Model 2	287.57	0.77
800 RPM–100 Mm/min	Experiment	247.956	–
	Model 1	231.101	7.2
	Model 2	248.565	0.24
1200 RPM–40 Mm/min	Experiment	304.592	–
	Model 1	273.401	11.4
	Model 2	305.529	0.30
1200 RPM–70 Mm/min	Experiment	295.828	–
	Model 1	278.609	6.1
	Model 2	290.034	1.99
1200 RPM–100 Mm/min	Experiment	296.551	–
	Model 1	237.61	24.8
	Model 2	288.301	2.86
1600 RPM–40 Mm/min	Experiment	357.146	–
	Model 1	284.403	25.5
	Model 2	365.044	2.16
1600 RPM–70 Mm/min	Experiment	338.173	–
	Model 1	295.15	14.5
	Model 2	344.461	1.85
1600 RPM–100 Mm/min	Experiment	308.893	–
	Model 1	258.503	19.4
	Model 2	318.005	2.86

The maximum error percentage in the model one was approximately 13.2% for the rotational velocity of 800 RPM and the transverse velocity of 40 mm/min, while after applying the calculated values in the model, it was decreased to 1.83%. Furthermore, the smallest percentage (0.24%) was obtained in the second model in which the rotational velocity was 800 RPM and the transverse velocity was 100 mm/min.

Meanwhile, for the rotational speed of 1200 RPM, the maximum error percentage value at the transverse speed of 40 mm/min was approximately 11.4%, however it decreased to 0.3%, after applying the calculated Young's modulus values in the second model. The percentage of the error in the second model lightly increased to 1.99 and 2.86% as the transverse speed increases from 70 mm/min to 100 mm/min.

As the rotational velocity rose to 1600 RPM, the percentage of the error in all specimens had also increased. Similar to the rotational speeds of 800 RPM and the 1200 RPM, the maximum error percentage was found in the transverse speed of 40 mm/min (about 25.5%), which was decreased to 2.16% in the second model. It was also detected that, after applying the temperature dependent values, the minimum value of the error percentage was observed in the transverse speed of 70 mm/min around 1.85% present. As the transverse speed increased to 100 mm/min, the percentage of the error also increase up to 2.86%. Consequently, the results of all models show that the increase in the rotational and transverse velocity increases the error percentage.

3.4 Temperature Results

As observed in Fig. 4, the temperature in the model one (with constant Young's Modulus) is always less than the experimental results. This could happen, because of the inaccuracy in the defined input parameter due to the lack of the material data. Conversely, more accurate results are contained in the second model where the temperature dependent Young's modulus values were used, and the gap between the predicted temperature values and the experimental measurements is decreased.

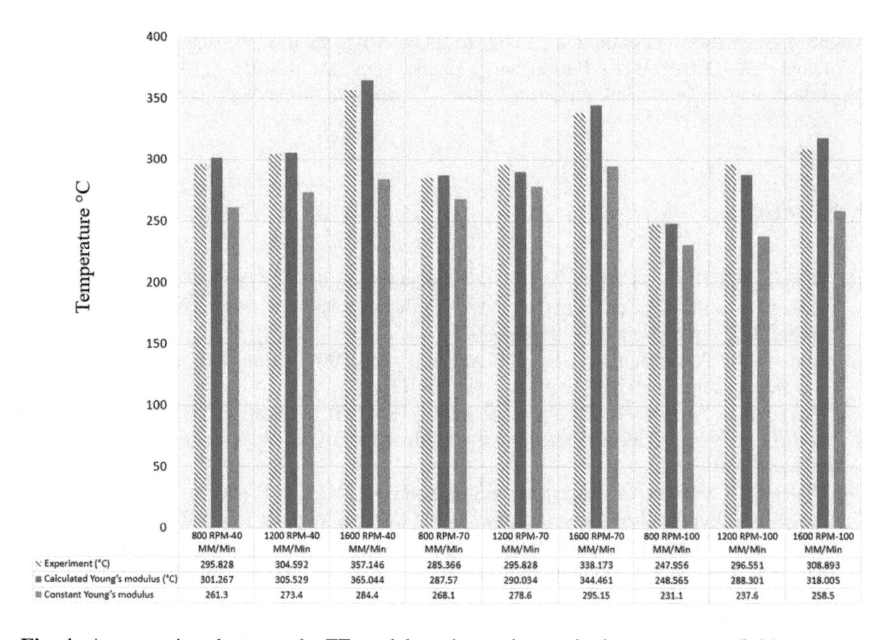

Fig. 4 A comparison between the FE models and experiments in the temperature field

4 Conclusions

The results presented showed that Young's modulus is one of the key properties of the material in FSW simulation and the temperature dependent values of the Young's modulus significantly reduce the gap between reality and the model simulation.

The following findings of this study are the summarize as below,

- This study has shown that the increase in temperature from 25 °C to the AA6061-T6 melting point 580 °C, results in the decrease of the Young's modulus in the range of 70–0.8 GPa.
- According to the experimental measurements, the peak temperature measured at the mid position of the weld (around 365 °C), at the transverse velocity of 40 mm/min and the rotational velocity of 1600 RPM.
- Compared to the experiments, the results of the first FE model showed that at the transverse velocity of 40 mm/min and the rotational velocity of 1600 RPM, the temperature was around 284 °C with the absolute error of 20.36%, while the temperature in the second model with the same process parameter was 365 °C with an absolute error of 2.21%. This issue confirms that the second model has a higher accuracy.
- The results also indicated that during all steps of the welding, the temperature is always lower than the melting temperature of 6061-T6 aluminium alloy which is at 580 °C.

Acknowledgements The authors would like to acknowledge the financial support from Universiti Teknologi PETRONAS (UTP), Bandar Seri Iskandar, Perak Darul Ridzuan, Malaysia. Moreover, the authors would like to thank Professor Wallace Kaufman for his endless support.

References

1. Dialami N, Cervera M, Chiumenti M, de Saracibar CA (2017) A fast and accurate two-stage strategy to evaluate the effect of the pin tool profile on metal flow, torque and forces in friction stir welding. Int J Mech Sci 122:215–227
2. Hodowany J, Ravichandran G, Rosakis A, Rosakis P (2000) Partition of plastic work into heat and stored energy in metals. Exp Mech 40:113–123
3. Meyghani B, Awang MB, Emamian SS, Mohd Nor MKB, Pedapati SR (2017) A comparison of different finite element methods in the thermal analysis of Friction Stir Welding (FSW). Metals 7:450
4. Dialami N, Chiumenti M, Cervera M, de Saracibar CA (2015) Challenges in thermo-mechanical analysis of friction stir welding processes. Arch Comput Methods Eng 1–37
5. Dialami N, Chiumenti M, Cervera M, Segatori A, Osikowicz W (2017) Enhanced friction model for Friction Stir Welding (FSW) analysis: simulation and experimental validation. Int J Mech Sci
6. El-Sayed MM, Shash AY, Mahmoud TS, Rabbou MA (2018) Effect of friction stir welding parameters on the peak temperature and the mechanical properties of aluminum alloy 5083-O. Improved performance of materials. Springer, Berlin, pp 11–25

7. Emamian S, Awang M, Yusof F, Hussain P, Mehrpouya M, Kakooei S et al (2017) A review of friction stir welding pin profile. In: Awang M (ed) 2nd international conference on mechanical, manufacturing and process plant engineering. Springer Singapore, Singapore, pp 1–18
8. Veljić DM, Rakin MP, Perović MM, Međo BI, Radaković ZJ, Todorović PM et al (2013) Heat generation during plunge stage in friction stir welding. Therm Sci 17:489–496
9. Sun T, Roy M, Strong D, Withers PJ, Prangnell PB (2017) Comparison of residual stress distributions in conventional and stationary shoulder high-strength aluminum alloy friction stir welds. J Mater Process Technol 242:92–100
10. Garg A, Bhattacharya A (2017) On lap shear strength of friction stir spot welded AA6061 alloy. J Manuf Processes 26:203–215
11. Awang M, Khan SR, Ghazanfar B, Latif FA (2014) Design, fabrication and testing of fixture for implementation of a new approach to incorporate tool tilting in friction stir welding. In: MATEC web of conferences, p 04020
12. Guerdoux S, Fourment L (2009) A 3D numerical simulation of different phases of friction stir welding. Model Simul Mater Sci Eng 17:075001
13. Bussetta P, Feulvarch É, Tongne A, Boman R, Bergheau J-M, Ponthot J-P (2016) Two 3D thermomechanical numerical models of friction stir welding processes with a trigonal pin. Numer Heat Transfer, Part A Appl 70:995–1008
14. Jain R, Pal SK, Singh SB (2016) Finite element simulation of temperature and strain distribution during friction stir welding of AA2024 aluminum alloy. J Inst Eng (India): Ser C, pp 1–7
15. Riahi M, Nazari H (2011) Analysis of transient temperature and residual thermal stresses in friction stir welding of aluminum alloy 6061-T6 via numerical simulation. Int J Adv Manuf Technol 55:143–152
16. Aziz SB, Dewan MW, Huggett DJ, Wahab MA, Okeil AM, Liao TW (2016) Impact of Friction Stir Welding (FSW) process parameters on thermal modeling and heat generation of aluminum alloy joints. Acta Metall Sin (Eng Lett) 29:869–883
17. Maisonnette D, Bardel D, Robin V, Nelias D, Suery M (2017) Mechanical behaviour at high temperature as induced during welding of a 6xxx series aluminium alloy. Int J Press Vessels Pip 149:55–65
18. Al-Badour F, Merah N, Shuaib A, Bazoune A (2013) Coupled Eulerian Lagrangian finite element modeling of friction stir welding processes. J Mater Process Technol 213:1433–1439
19. Meyghani B, Awang M, Emamian S (2016) A comparative study of finite element analysis for friction stir welding application. ARPN J Eng Appl Sci 11:12984–12989
20. Meyghani B, Awang M, Emamian S, Khalid NM (2017) Developing a finite element model for thermal analysis of friction stir welding by calculating temperature dependent friction coefficient. In: Awang M (ed) 2nd international conference on mechanical, manufacturing and process plant engineering. Springer Singapore, Singapore, pp 107–126
21. Emamian S, Awang M, Hussai P, Meyghani B, Zafar A (2006) Influences of tool pin profile on the friction stir welding of AA6061
22. Meyghani B, Awang M, Emamian S (2017) A mathematical formulation for calculating temperature dependent friction coefficient values: application in Friction Stir Welding (FSW). In: Defect and diffusion forum, pp 73–82

The Effect of Pin Profiles and Process Parameters on Temperature and Tensile Strength in Friction Stir Welding of AL6061 Alloy

S. Emamian, M. Awang, F. Yusof, Patthi Hussain, Bahman Meyghani and Adeel Zafar

Abstract The main source of the heat generation during the Friction Stir Welding (FSW) is the friction force between tool and workpiece and the plastic deformation. The geometry of the tool including the pin and the shoulder highly affects the friction force. In this study, the effects of different pin profiles with different rotational and traversing speed are evaluated in order to obtain the optimum pin profile using heat generation and tensile strength. Three different rotational speed and welding speeds are applied with threaded cylindrical, conical, stepped conical and square pin profiles. Thermocouples K type have been embedded in order to record the temperature during the welding at the advancing and the retreating side. Moreover, tensile test and microstructure analysis are performed in order to study the microstructure. The results of experimental process and design of experiments are correlated well. The better joint produced with threaded cylindrical tool pin profile with rotation speed of 1600 rpm and welding speed of 40 mm/min.

Keywords FSW · Pin profile · Friction stir welding · Heat generation · Tensile strength

1 Introduction

Friction Stir Welding (FSW) was invented and patented by Thomas et al. at The Welding Institute (TWI) [1, 2]. There are three stages in the process plunging; welding stage and plunging out step. In the plunge stage, FSW tool which is made up of a pin and a shoulder, penetrates the plates. In some cases, there is a dwell time in which

S. Emamian · M. Awang (✉) · P. Hussain · B. Meyghani · A. Zafar
Department of Mechanical Engineering, Universiti Teknologi PETRONAS, Bandar Seri Iskandar, Malaysia
e-mail: mokhtar_awang@utp.edu.my

F. Yusof
Department of Mechanical and Manufacturing, University of Malaya, Kuala Lumpur, Malaysia

M. Awang (ed.), *The Advances in Joining Technology*, Lecture Notes in Mechanical Engineering, https://doi.org/10.1007/978-981-10-9041-7_2

the rotating tool remains at the plate without forward motion. In the welding stage, the rotating tool moves forward along the welding seam in order to form a weld bead [3, 4]. The schematic of FSW is shown in Fig. 1 [5].

Although all of the welding stages are significant, the significance of the plunging stage is more than other stages, because the main part of the heat will be generated at this stage and the workpiece will be affected extremely by high temperature and stresses [6]. It should be noted that most tool wear occurs during plunge stage due to the high load and flow stress [7, 8]. In FSW, the parameters that influence quality of joint and reduce tool wear are rotational speed and traverse speed while it was reported that the influence of the geometry of the tool is more than the abovementioned process parameters [9, 10]. The geometry of the tool is separated into two different parts of shoulder and pin whereby both have a significant effect on the material flow and the welding temperature. Among all shapes that scholars considered, some of them are not compared with each other in the literatures [11–21]. To illustrate, Patil and Soman [13] only utilized Tri-flute and taper screw during different welding speeds. In the same way, [16] investigated different pin profiles in which the square profile was absent. [22] studied the influence of the pin profiles and the shoulder diameter on the formation of FSW stirring zone. They found that the square pin profile with 18 mm shoulder diameter have better weld quality in comparison with other pin profiles. In other research, they investigated the influence of the different pin profiles and the welding speed during the FSW. Their results indicated that the square pin profile produced defect free weld compared to other pin profiles [12].

[23], studied the influence of different pin profiles on the metallurgical and mechanical properties of Al-Metal Matrix Composite. They concluded that the square pin profile has better tensile strength in comparison with other pin profiles. However, the threaded cylindrical was not involved in their study. [24], investigated the effect of the different pin profiles on the mechanical properties of FSW of pure copper. Their results showed that the square pin profile have higher mechanical properties and better grain structure in comparison with the threaded cylindrical.

There are many considerable experimental and numerical studies on FSW of different alloys. [25] comprehensively reviewed the friction stir processes. [26] reviewed

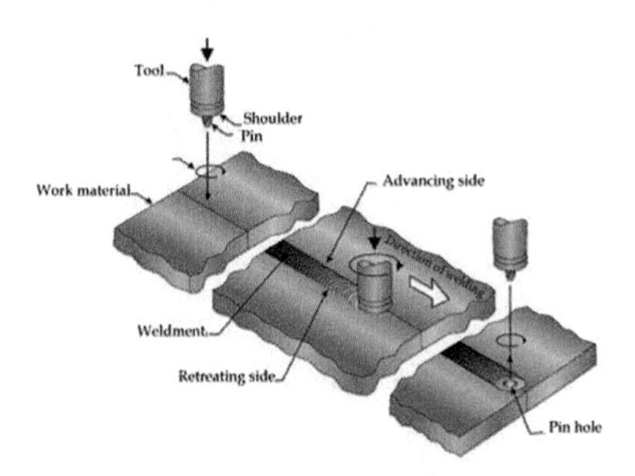

Fig. 1 Schematic of friction stir welding [3]

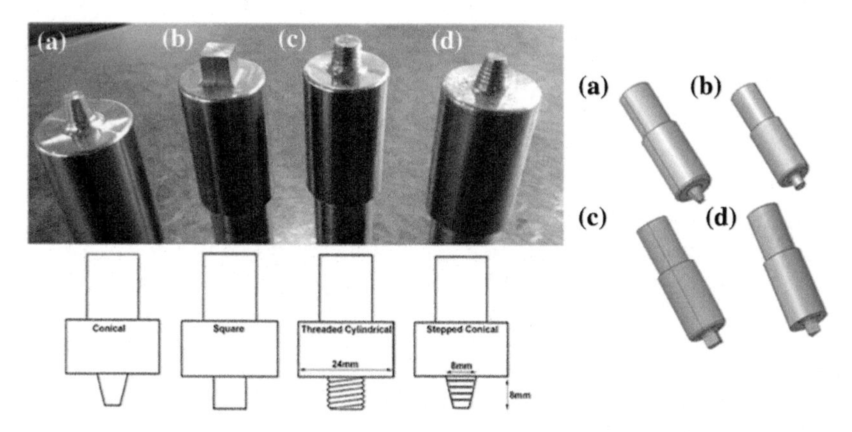

Fig. 2 Different pin profiles. **a** Conical. **b** Square. **c** Threaded cylindrical. **d** Stepped conical

the structure and the properties of FSW. [27] gave a broad review of numerical analysis of FSW. [28] reviewed the different pin profiles. [29] studied finite element modeling approach for Friction Stir Spot Welding (FSSW) on Al6061. They applied adaptive mesh to reduce the high distortion during simulation. Besides, [30] considered the plunge stage using numerical modeling and experimental. They utilized Al2024 alloy for the experiment.

The critical part of FSW is the pin profile, because it affects the welding quality. Therefore, regarding the review of literature, in this paper, four FSW tool pin profile have been selected including; square (s), threaded cylindrical (TC), stepped conical (SC), and conical (C). Figure 2 shows the schematic of profiles.

An appropriate design of the tool (especially the pin profiles) is able to generate proper heat and mixing the plasticized materials. Another significant factor which affects the heat generation is the process parameters such as rotating speed, traveling speed [25]. In this paper, in order to optimize the pin profile of FSW tool and process parameter during FSW, the thermomechanical behavior of the welded samples has been studied in detail.

2 Methodology

2.1 Experimental Set up

As mentioned earlier, four different pin profiles were selected for experiments. The heat treated H13 steel was used as the tool material and the workpiece AA6061-T6 with the dimension of 100 mm × 100 mm × 10 mm has been clamped. The chemical composition of a workpiece and the tool is shown in Table 1.

Table 1 Chemical composition and mechanical properties of base metal and the tool

AA6061-T6	**Chemical composition %—workpiece**								
	Mg	Cr	Ti	Zn	Mn	Fe	Si	Cu	Al
	0.98	0.19	0.05	0.01	0.07	0.3	0.47	0.23	Balanced
	Mechanical properties								
	UTS (Mpa)	Yield Strength (Mpa)	Elongation (%)						
	305	253	12						
H13	**Chemical composition %—tool**								
	C	Si	Mn	Cr	Mo	V	P	S	Fe
	0.4	0.92	0.34	5.07	1.25	0.95	0.019	0.001	Balanced

The experiments were performed by FSW-TS-F16 friction stir welding device host machine. The single pass welding procedure has been used to fabricate the joints. The heat treatment operation is done on the FSW tool. The first step of the heat treatment is preheating cycle. According to the standard, ASTM E8 temperature for preheating is around 760 °C, held for 15 min. Then, it is soaking cycle in the austenite formation zone in which the temperature is raised up to 1010 °C and held for 30 min. After the soaking cycle, the tool is removed from the furnace and cooled to 65 °C, when the furnace has reached a temperature of 565 °C, the tools are allowed to a temper for 2 h [31]. Three rotational speeds and three traverse speeds were selected to evaluate the temperature during different speeds that are listed in Table 2. In order to record the temperature during the welding, thermocouples (K type) are embedded in the advancing side and retreating sides with the specified distance. Figure 3 shows the position of the thermocouples.

Table 2 Welding parameters and tool dimension

Process parameters	Values
Rotational speed (rpm)	800, 1200, 1600
Traveling speed (mm/min)	40, 70, 100
Axial force (kN)	7
Pin length (mm)	8
Tool shoulder diameter (mm)	24
Pin diameter (mm)	8

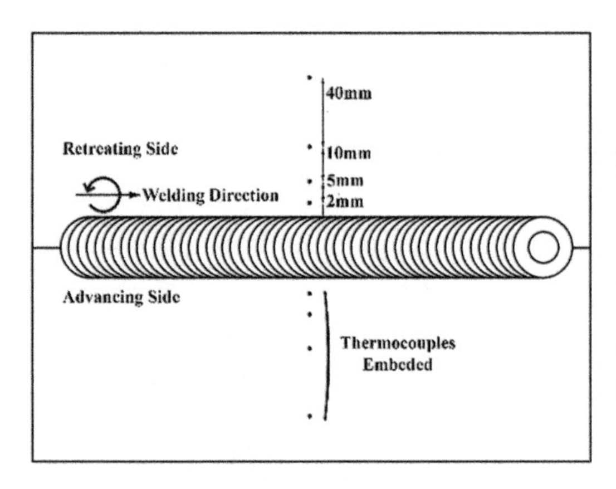

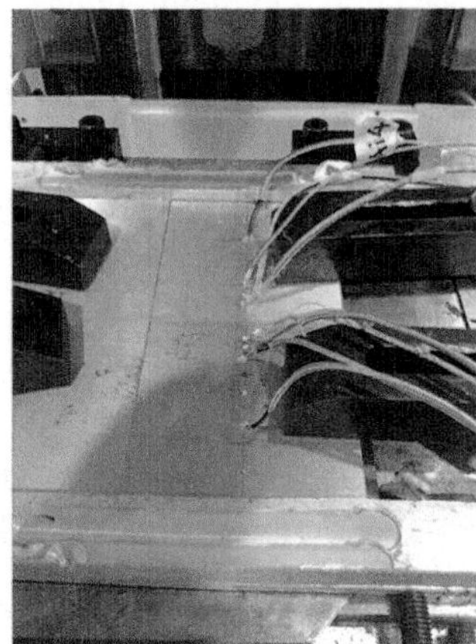

Fig. 3 Workpiece and thermocouples position

2.2 *Microstructure Analysis*

Microstructure analysis of FSW joints is performed as per ASTM to optimize the tool pin profile. In order to cut the samples, a wire cut machine has been used.

Then, the samples are grinded to remove material deformed produced from sectioning. Grinding process is followed by the polishing process. Polishing is removing the scratches from the surface of specimens to prepare them for the etching process. In order to perform etching, the specimens are immersed into Keller's reagent for 10–20 s. The chemical composition of Keller's reagent are listed in Table 3 [32, 33].

2.3 *Tensile Test*

Tensile test is performed using a universal testing machine 50KN Amsler HA50 with standard ASTM: E8/E8M with constant crosshead speed of 0.9 mm/min. Tensile testing is a fundamental material science test that is subjected to a controlled tension force until the fracture. The results from the test are commonly used to select a material for an application, to control the quality of the weld, and also to predict the fraction of the material under different types of forces. Properties that are directly measured during a tensile test are ultimate tensile strength, maximum elongation and reduction in area. Tensile test performed using a universal testing machine with standard ASTM: E8/E8M utilized for tensile and performed with constant crosshead speed of 0.9 mm/min. Figure 4 shows the schematic of tensile samples for the test according to ASTM E8.

Table 3 Chemical composition of Keller's reagent

Keller's reagent				
Composition	HF	HCl	HNO_3	Water
Volume (ml)	2	3	5	190

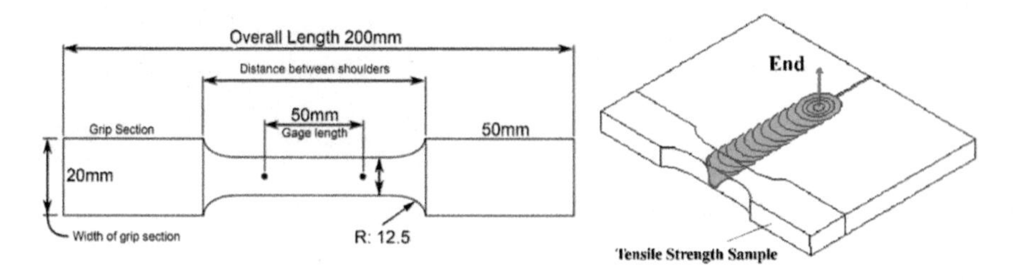

Fig. 4 Schematic of tensile test sample and dimension

2.4 Design of Experiment (DOE)

Design of experiments (DOE) is a systematic method in order to determine the relationship between factors affecting in a process and also to find the cause-and-effect relationships. Basically, DOE is a statistical technique for analyzing and organizing the experiments [34]. In DOE, the factors comprise of different parameters which are controlled by researchers, meanwhile the responses make up the dependent variable, which in this case, refers to productivity. To implement a DOE technique, some steps need to be followed such as: choosing the factors and their levels, choosing a response variable, choosing the experimental design, performing the experiments, analyzing the data, and promoting the best option [35].

In this research, the general factorial design has been selected for evaluating the effect of the several parameters on the heat generation and peak stress. In the first step for conducting the results. Table 4 and Table 5 show the factors and their levels for the heat generation and the stress respectively.

Table 4 Factors and levels for heat generation

Factors	Levels			
Pin profile (A)	1	2	3	4
	Square	Threaded cylindrical	Stepped conical	Conical
Rotational speed—rpm (B)	800, 1200, 1600			
Welding speed—mm/min (C)	40, 70, 100			

Table 5 Factors and levels for peak stress

Factors	Levels			
Pin profile (A)	1	2	3	4
	Square	Threaded cylindrical	Stepped conical	Conical
Rotational speed—rpm (B)	800, 1600			
Welding speed—mm/min (C)	40, 100			

3 Results and Discussion

3.1 Heat Generation

Investigation of the heat generation in FSW is a complicated phenomenon and needs to be compared with the experimental data. In this study, heat generation of different parameters such as rotational speed, welding speed and tool pin profiles are investigated. Figure 5 illustrates the results of heat generation during FSW for advancing and retreating sides. These histograms show the maximum temperature from eight thermocouples for 36 sample welds. Highest temperature obtained from rotation speed of 1600 rpm for all pin profiles. For instance, for the threaded cylindrical tool pin profile, as the welding rotation speed increases from 800 to 1600 rpm, the temperature increases from 247.95 to 357.14 °C.

The Figure shows the relation of different parameters. As the welding increases from 40 mm/min to 100 mm/min, the temperature drops from 357.14 to 247.95 °C. This happens due to the reduced heat input per unit length and dissipation of heat over a wider region of workpiece at higher welding speed. On the other hand, by increasing the rotation speed, temperature would be increased. As the welding rotation speed increases from 800 to 1600 rpm, the temperature increases from 203 to 459.6 °C for the threaded cylindrical tool pin profile in the advancing side. This happens due to friction between tool and workpiece that generate more heat. Moreover, surface area of the tool effects heat generation. In this study, square pin profile produced a higher temperature around 460 °C due to its surface shape that creates more friction during FSW.

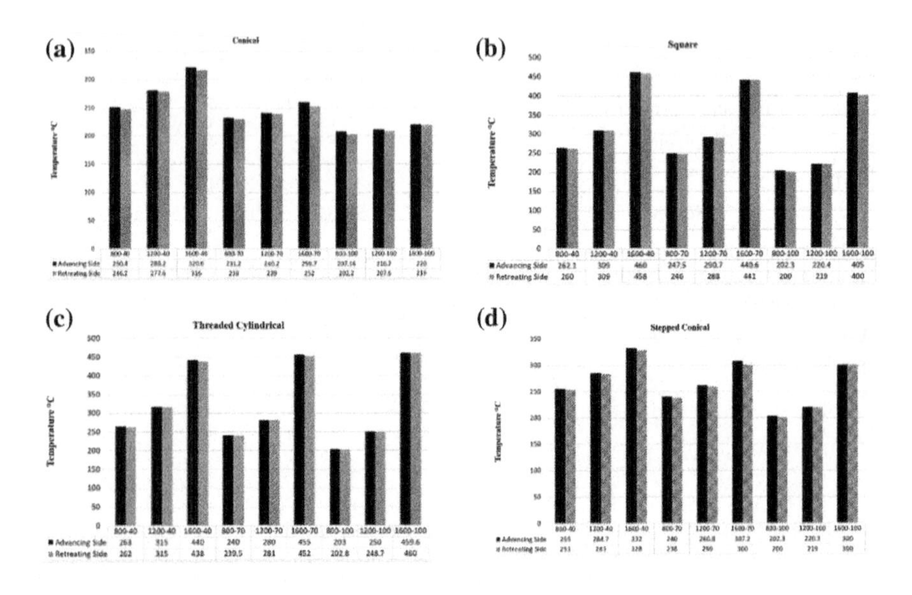

Fig. 5 Maximum temperature for different pin profiles

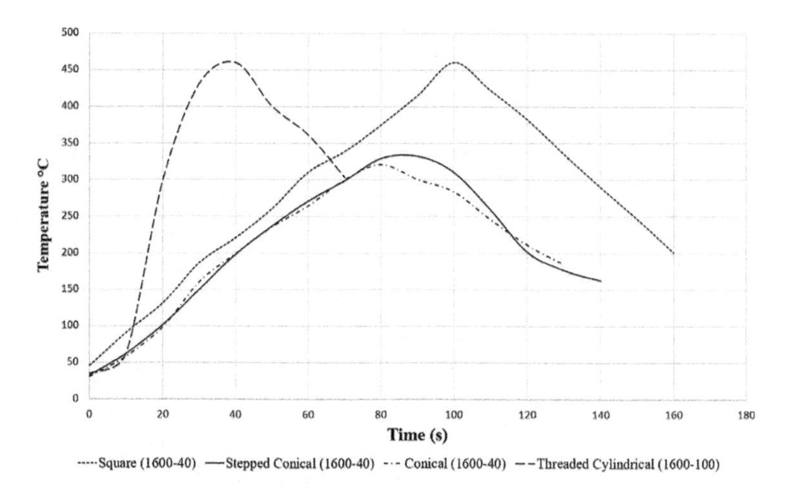

Fig. 6 Temperature graphs from different pin profiles

Advancing side in FSW process is the location from where the solid material starts to transform into semi-solid one and flows around the tool pin plunged into the material. The semi-solid material retreated and cooled in the retreating side. Therefore, advancing side has more solid state nature at any point of time/location compare to retreating side during FSW process. Therefore, advancing side should generate higher friction stress (unbalanced frictional force) which ultimately generates more heat and raises the peak temperature. Moreover, advancing side produced higher temperatures in comparison with retreating side due to pushing the material at the first connection and forwarding it to retreating side.

Temperature under the shoulder is higher due to high energy density. The peak of the temperature is about 80% of the material melting point 582–652 °C. From 36 welded samples, those where generates highest temperatures compared with each other to analyze heat generation during FSW with different pin profiles. Figure 6 shows the combined graphs. As can be seen in the Figure, square pin profile produced a higher temperature around 450 °C due to its surface shape that creates more friction during FSW.

During the welding process, the shoulder was the same, but pin profiles changed with different speeds. We could obtain different temperatures from different speeds. However, the graphs show that differences between peak temperatures of samples welded by different pin profiles are very little and not significant. According to Eqs. (1) and (2), pin profile affects heat generation.

$$Q_1 = 2/3\,\pi\,\tau_{contact}\omega\left(R^3_{shoulder} - R^3_{pin}\right)(1 + tan_{\alpha}) \tag{1}$$

$$Q_2 = 2\,\pi\,\tau_{contact}\omega R^2_{pin} H_{pin} \tag{2}$$

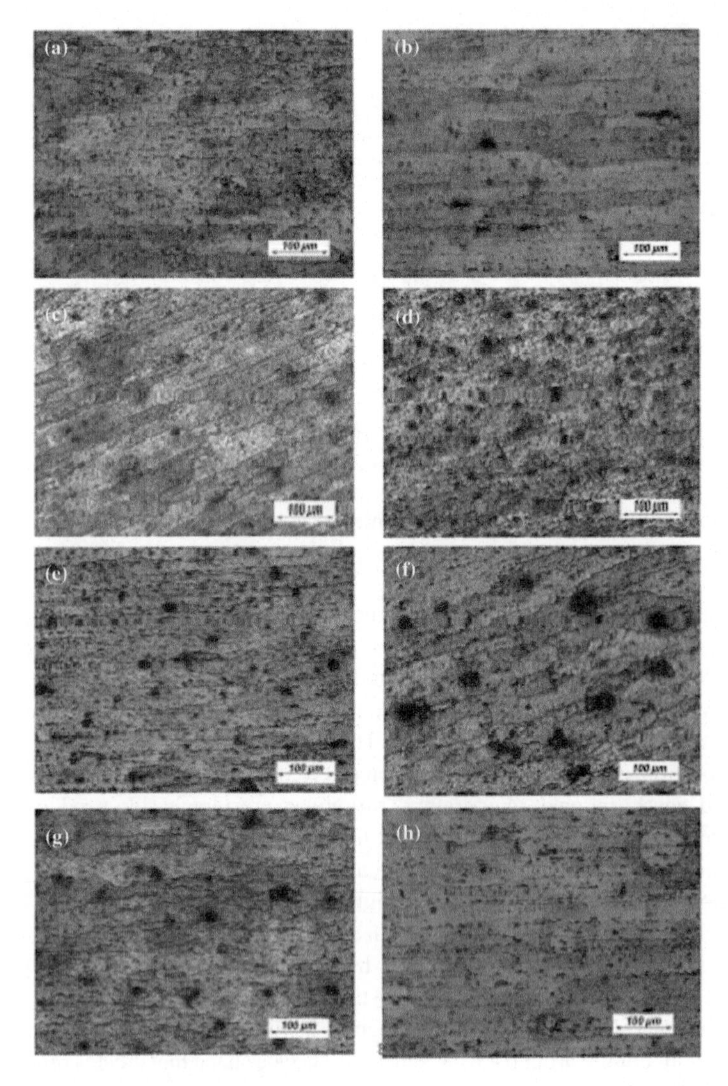

Fig. 7 Microstructure of advancing and retreating side. **a** Advancing side of stepped conical pin profile. **b** Advancing side of conical pin profile. **c** Advancing side of threaded cylindrical pin profile with rotation speed of 1600 rpm and welding speed of 100 mm/min. **d** Advancing side of threaded cylindrical pin profile with rotation speed of 800 rpm and welding speed of 40 mm/min. **e** Retreating side of stepped conical. **f** Retreating side of conical pin profile. **g** Retreating side of threaded cylindrical with rotation speed of 1600 rpm and welding speed of 100 mm/min. **h** Retreating side of threaded cylindrical with rotation speed of 800 rpm and welding speed of 40 mm/min

Equations (3) and (4) give the local heat generation rate due to friction and deformation work respectively.

$$de_f = \delta\,(r\omega - U\,sin\,\theta)\,\mu_f\,p dA \tag{3}$$

$$de_s = (1 - \delta)\,(r\omega - U\,sin\,\theta)\,\tau_y dA \tag{4}$$

where δ is the extent of the slip, μ_f is the friction coefficient, μ_s is the shear yield stress and p is the local pressure applied by the tool on the elemental area dA. When δ is 1, no material sticks to the tool and all the heat is generated by friction. In contrast, when $\delta = 0$, work-piece material sticks to the tool and all the heat is generated by plastic deformation [36, 37].

According to the results of the experiment, higher temperature is generated from higher rotational speed and lower temperature comes from lower traverse speed. Therefore, increasing the rotational speed will increase the temperature and increasing the welding speed will reduce temperature that proved with other researches that studied in literature review. Moreover, the highest temperature is from a square pin profile that can be demonstrated with its surface area connection with the materials.

3.2 Microstructure Analysis

The etched specimens are examined using optical microscope Leica MC170 HD and the grain size of the cross section of the joints are analyzed. The cut specimens are immersed into Keller's reagent for 10–20 s for the etching process. Finally, Scanning Electron Microscope (SEM) is used in order to investigate the microstructure.

It is notable that all microstructures of threaded cylindrical are homogenous and there are no significant differences between advancing and retreating side. Due to utilizing threaded shapes during FSW mixing of the materials is higher and also the grain size is better. The role of the pin profile for making the microstructure is not inevitable. The results showed that, the threaded cylindrical made a homogeneous microstructure in FSW in comparing with conical and stepped conical profiles. During FSW, the temperature variation is very high, therefore, the recrystallization of samples is not complete and are broken or are in the shape of a pancake. As per discussion earlier about the heat generation by increasing the rotational speed, the temperature is also increased and thus, the microstructure should be coarse. On the other hand, by increasing the welding transverse speed, the temperature falls down and thus the microstructure is finer.

A comparison between different microstructures at the stir zones in advancing and retreating sides under the various FSW conditions is shown in Fig. 7a–h. Figures (a) and (b) are related to advancing and retreating side of the conical profile with rotation speed of 1600 rpm and welding speed of 40 mm/min. Figures (c) and (d) are microstructure of square profiles. Threaded cylindrical shown in figures (e) and (f) and stepped conical is shown with figures (g) and (h). The results indicated that the

Table 6 Grain size of joints in various samples

Pin profile	Rotational speed (rpm)	Welding speed (mm/min)	Grain size (μm)	
			AS	RS
Stepped conical	800	100	38.9	34.7
Conical	800	40	49.3	43.3
Threaded cylindrical	1600	100	32.26	31.8
Threaded cylindrical	800	40	39.8	33

microstructure of the sample with rotational speed of 1600 rpm with 100 mm/min welding speed is finer.

As per discussion earlier about heat generation by increasing rotational speed, temperature would be increased. Therefore, the microstructure should be coarse. On the other hand, by increasing the welding speed, temperature falls down and microstructure would be coarse again. Table 6 indicates the grain sizes.

The role of pin profile for making the microstructure is not inevitable. Threaded cylindrical made a homogeneous microstructure in FSW in comparing with conical and stepped conical profiles.

In comparing with mechanical test and heat generation, sample with rotational speed of 1600 rpm with 100 mm/min welding speed produced sounds joint with finer microstructure.

3.3 *Tensile Strength Analysis*

After analyzing the microstructure, as can be seen in Fig. 8 some of the samples produced wormholes, therefore among 36 samples only 11 samples are Defect-free, and were selected for the tensile test. As mentioned earlier, these samples were tested using a universal testing machine with standard ASTM: E8/E8M utilized for tensile and performed with constant crosshead speed of 0.9 mm/min. Transverse tensile properties of FSW joints such as yield strength, tensile strength, and percentage of elongation have been evaluated. Table 7 illustrates the mechanical properties of selected samples. In another aspect, primary parameters such as welding speed has a significant role in tensile properties.

As shown, maximum tensile strength (73.91% of parent material) is observed in that specimen in which the threaded cylindrical profile with rotational speed of 1600 rpm and welding speed of 100 mm/min is used. In contrast, the inferior properties belong to conical pin profile with the rotational speed of 800 rpm and the transverse speed of 40 mm/min. Therefore, lower tool rotational speeds produced insufficient intermixing. Thus, the mechanical strength will decrease at lower rota-

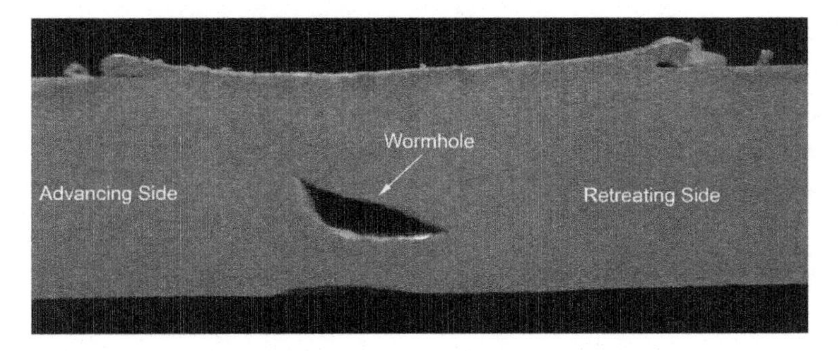

Fig. 8 Wormhole in FSW samples

Table 7 Tensile properties of 11 samples

Samples	0.2% yield strength	Tensile strength (MPa)	Elongation %
Conical (800-40)	113.50	180.45	6.25
Stepped conical (800-100)	129.70	188.21	5.10
Threaded cylindrical (800-40)	112.02	200.70	8.6
Threaded cylindrical (800-70)	125.63	212.22	7.66
Threaded cylindrical (800-100)	130.76	213.17	5.14
Threaded cylindrical (1200-40)	112.30	201.23	8.84
Threaded cylindrical (1200-70)	124.32	210.04	7.87
Threaded cylindrical (1200-100)	131.50	193.84	3.88
Threaded cylindrical (1600-40)	109.91	186.39	5.85
Threaded cylindrical (1600-70)	121.32	208.21	8.67
Threaded cylindrical (1600-100)	132.77	230.359	5.78

tional speeds [38]. Moreover, it is observed that by increasing the welding speed, tensile strength will be increased. In another aspect, primary parameters such as welding speed have a significant role in the tensile properties.

Table 8 shows the tensile properties of those specimens in which maximum heat is generated. These results showed that increasing the rotation speed can increase temperature. On the other hand, by reducing the welding transverse speed temperature is also increased.

Table 8 Tensile properties of FSW joints with maximum heat generation

Samples	0.2% yield strength	Tensile strength (MPa)	Elongation %
Conical (1600-40)	113.50	141.246	5.68
Square (1600-40)	129.70	104.841	2.213
Threaded cylindrical (1600-100)	132.77	230.359	5.78
Stepped conical (1600-40)	125.63	140.458	7.66

3.4 Design of Experiments Analyzes

3.4.1 Statistical Analysis for Heat Generation and Peak Stress

After choosing the factors, levels and the experimental parameters, data collection needs to be conducted by doing the experiments. As mentioned earlier by using general factorial design method, 36 conditions for the heat generation and 16 experiments for the peak stress are optimized. In order to decrease potential errors, two replicates for the heat generation and the peak stress is assumed. Therefore, as can be seen in Table 9 the numbers of the experiments for heat generation is equal to 72 (The number of experiments = 4 * 3 * 3 * (2 replicates) = 72).

As illustrated in Table 10, the number of the experiments for peak stress is equal to = 4 * 2 * 2 * (2 replicates) = 36.

3.4.2 Determining of Significant Factors for Heat Generation

In this study, Expert-design software is used for conducting statistical analyses. The result of ANOVA for the heat generation and the peak stress are presented in Table 11. Basically, *P*-value is a significant parameter that is usually used to identify the statistically significant factors due to it has influenced the final responses. It should be noted that, whenever *P*-values are less than 0.05, factors should be considered as significant. In contrast, when *P*-values are more than 0.05, it should be assumed as an insignificant factor [39]. Based on Table 5, the Model, F-value of 6526.42 implies that the model is significant. In addition, the *P*-values is less than 0.0500 which indicate the model terms are significant. In this case A, B, C, AB, AC, BC, ABC are significant model terms.

Table 12 verifies the accuracy of the model. As can be seen in the table, The "Pred R-Squared" of 0.9994 is in reasonable agreement with the "Adj R-Squared" of 0.9997. "Adeq Precision" measures the signal to noise ratio that a ratio greater than 4 is desirable. So ratio of 262.871 indicates an adequate signal.

Table 9 Result of Experiments for heat generation

Run	Pin profile (A)	Rotational speed—rpm (B)	Welding speed—mm/min (C)	Heat generation (°C)	
				Advancing side	Retreating side
1	Square	800	40	262.1	260
2	Threaded cylindrical	800	40	263	262
3	Stepped conical	800	40	255	253
4	Conical	800	40	250.4	246.2
5	Square	1200	40	309	309
6	Threaded cylindrical	1200	40	315	315
7	Stepped conical	1200	40	284.7	283
8	Conical	1200	40	280.2	277.6
9	Square	1600	40	460	458
10	Threaded cylindrical	1600	40	440	438
11	Stepped conical	1600	40	332	328
12	Conical	1600	40	320.6	316
13	Square	800	70	247.5	246
14	Threaded cylindrical	800	70	240	239.5
15	Stepped conical	800	70	240	238
16	Conical	800	70	231.2	230
17	Square	1200	70	290.7	288
18	Threaded cylindrical	1200	70	280	281
19	Stepped conical	1200	70	260.8	259
20	Conical	1200	70	240.2	239
21	Square	1600	70	440.6	441
22	Threaded cylindrical	1600	70	455	452
23	Stepped conical	1600	70	307.2	300
24	Conical	1600	70	259.7	252
25	Square	800	100	202.3	200
26	Threaded cylindrical	800	100	203	202.8
27	Stepped conical	800	100	202.3	200
28	Conical	800	100	207.16	202.2
29	Square	1200	100	220.4	219
30	Threaded cylindrical	1200	100	250	248.7
31	Stepped conical	1200	100	220.3	219
32	Conical	1200	100	210.7	207.6
33	Square	1600	100	405	400
34	Threaded cylindrical	1600	100	459.6	460
35	Stepped conical	1600	100	300	300
36	Conical	1600	100	220	219

Table 10 Result of experiment for peak stress

Run	Pin profile (A)	Rotational speed—rpm (B)	Welding speed—mm/min (C)	Peak stress (MPa)	
				Advancing side	Retreating side
1	Conical	800	40	121.414	135
2	Stepped conical	800	40	178.561	160
3	Square	800	40	117.496	123
4	Threaded cylindrical	800	40	200.709	213
5	Conical	1600	40	141.246	161
6	Stepped conical	1600	40	140.458	149
7	Square	1600	40	104.841	123
8	Threaded cylindrical	1600	40	186.392	196
9	Conical	800	100	170.94	176
10	Stepped conical	800	100	188.218	225
11	Square	800	100	153.274	143
12	Threaded cylindrical	800	100	213.73	230
13	Conical	1600	100	180.456	173
14	Stepped conical	1600	100	131.943	138
15	Square	1600	100	136.556	125
16	Threaded cylindrical	1600	100	230.359	245

Table 11 ANOVA result for heat generation

Source	Sum of square	DF	Mean square	F value	*P*-value	Remarks
Block	14.07	1	14.07			
Model	4.376E+005	35	12502.97	6526.42	<0.0001	Significant
A	76208.97	3	25402.99	13260.09	<0.0001	Significant
B	2.313E+005	2	1.157E+005	60378.54	<0.0001	Significant
C	42508.87	2	21254.43	11094.59	<0.0001	Significant
AB	80365.51	6	13394.25	6991.66	<0.0001	Significant
AC	2262.53	6	377.09	196.84	<0.0001	Significant
BC	1475.16	4	368.79	192.50	<0.0001	Significant
ABC	3328.41	12	277.37	144.78	<0.0001	Significant
Residual	67.05	35	1.92			
Lack of Fit	67.05	34	1.97			
Pure Error	0.000	1	0.000			
Cor Total	4.377E+005	71				

Table 12 Model accuracy

Std. Dev.	1.38	R-squared	0.9998
Mean	286.88	Adj R-squared	0.9997
C.V.	0.48	Pred R-squared	0.9994
Press	284.21	Adeq precision	262.871

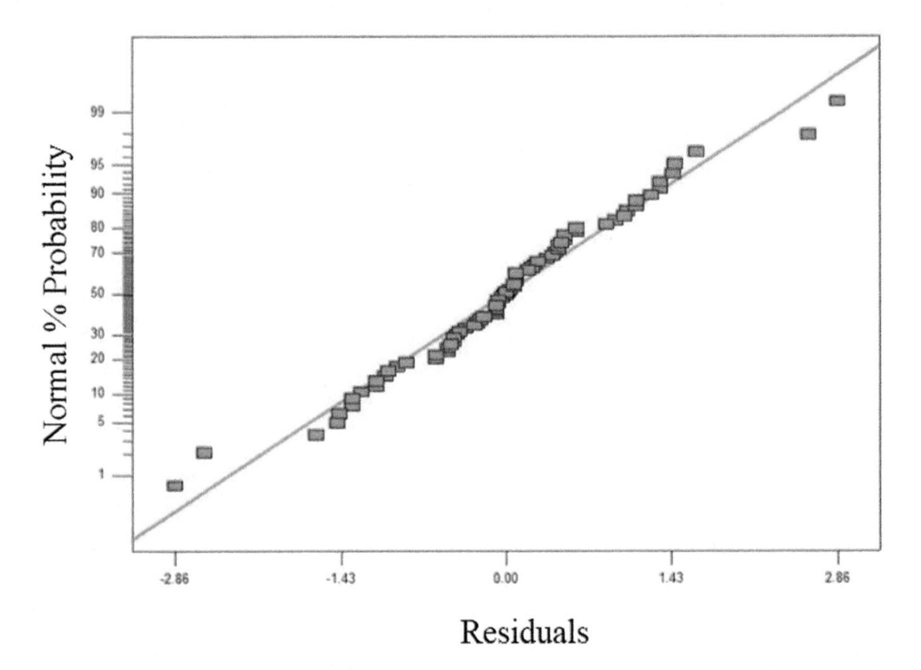

Fig. 9 Normal plot of residual

3.4.3 Residual Analysis

The residual versus predicted value and normal probability plots are two significant graphical approaches that were used in order to check the validity of the regression model [40]. The results of the residual versus predicted value shows that the difference between the predicted values and the observed values. If the residuals have a regular pattern, it will infer that the suggested model is not adequate. Moreover, residuals in normal probability plot should be laid in a straight line [41]. As can be seen in Fig. 9, the straight line confirms that the model is adequate and correct. Moreover, the structure less pattern of the residual versus predicted value confirms that the developed model is adequate (Fig. 10).

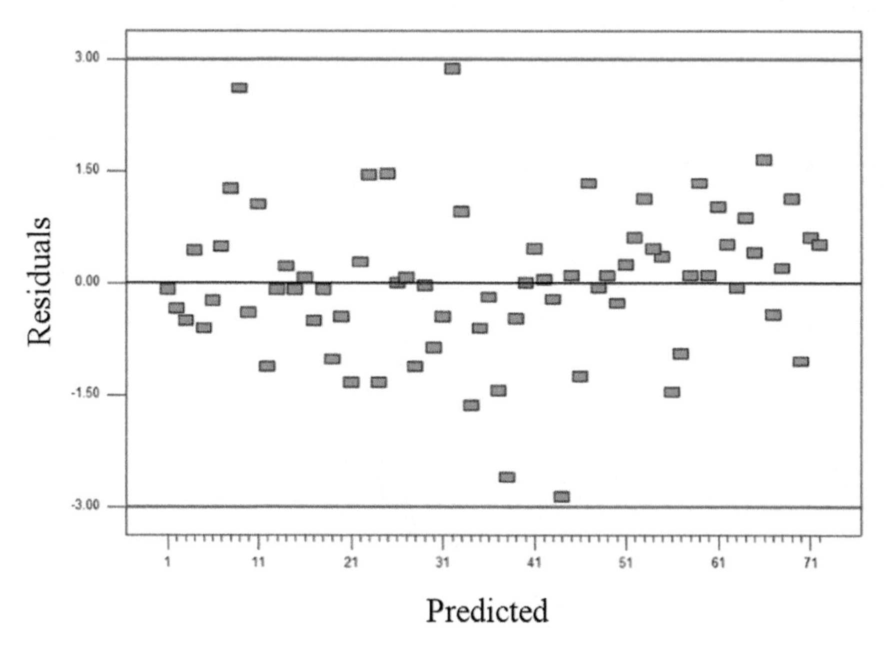

Fig. 10 Residual versus predicted value

3.4.4 Peak Stress

Determining of Significant Factors

Table 13 shows the result of ANOVA for the peak stress. As can be seen, factors A, B, C and interactions AB and ABC are significant and have a considerable effects on the peak stress.

Tabel 14 indicates that the "Pred R-Squared" of 0.8533 is in reasonable agreement with the "Adj R-Squared" of 0.9355. Also "Adeq Precision" measures the signal to noise ratio that a ratio greater than 4 is desirable. So ratio of 18.276 reveals an adequate signal.

3.4.5 Residual Analysis for Peak Stress

As can be seen in Fig. 11, the straight line confirms that the model is adequate and correct. Moreover, the structure less pattern of the residual versus predicted value confirms that the developed model is adequate and has a constant error (Fig. 12).

Table 13 ANOVA result for peak stress

Source	Sum of square	DF	Mean square	F value	Prob>F	Remarks
Block	438.13	1	438.13			
Model	43647.15	15	2909.81	30.01	< 0.0001	Significant
A	30756.93	3	10252.31	105.75	< 0.0001	Significant
B	1093.85	1	1093.85	11.28	0.0043	Significant
C	5236.71	1	5236.71	54.02	< 0.0001	Significant
AB	4155.39	3	1385.13	14.29	0.0001	Significant
AC	546.39	3	182.13	1.88	0.1766	Notsignificant
BC	267.99	1	267.99	2.76	0.1171	Notsignificant
ABC	1589.90	3	529.97	5.47	0.0097	Significant
Residual	1454.20	15	96.95			
Cor total	45539.48	31				

Table 14 Model accuracy for peak stress

Std. Dev.	9.85	R-squared	0.9678
Mean	165.99	Adj R-squared	0.9355
C.V.	5.93	Pred R-squared	0.8533
Press	6618.24	Adeq precision	18.276

3.4.6 Optimum Condition

The results of experiment and design of experiments have a good correlation. In most samples, advancing side produced higher temperature. Higher temperature obtained from square pin profile as discussed earlier due to its shape and geometry of contact area. However, from the experiments, we found that threaded cylindrical produced higher quality joints. It is observed that in some cases that temperatures are not exactly the same as DOE results because of some technical errors while getting temperature from thermocouples. Moreover, the results indicated that the highest tensile strength comes from threaded cylindrical with rotation speed of 1600 rpm and welding speed of 100 mm/min.

For the heat generation, the actual factor for the welding speed is obtained at 100 mm/min. Table 15 indicates the optimum parameters for heat generation and peak stress. The maximum temperature was 459.8 °C for the threaded cylindrical and rotation speed of 1600 rpm. Moreover, the stress of 237.68 MPa is obtained for threaded cylindrical.

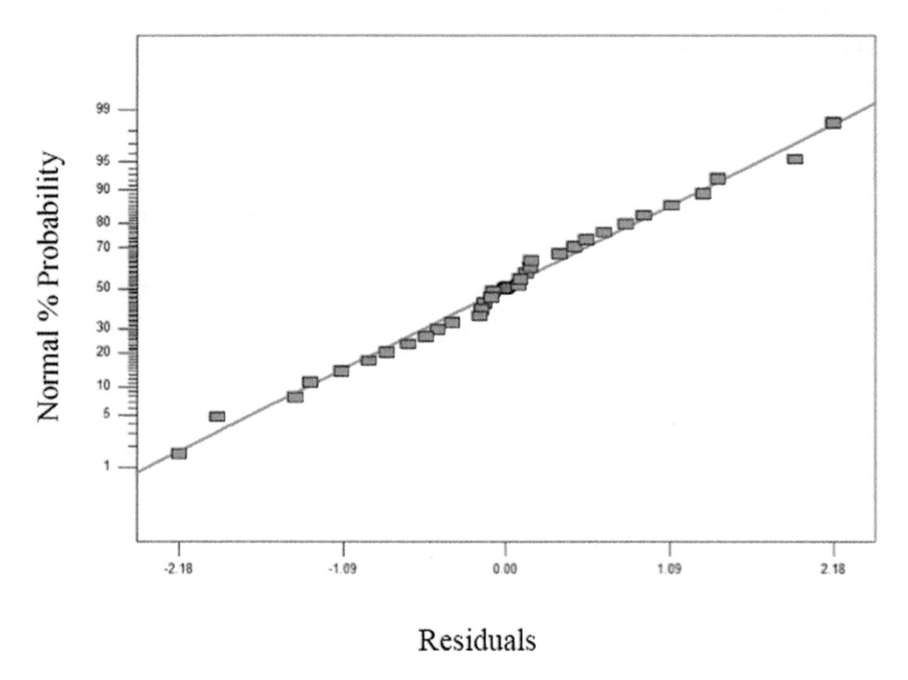

Fig. 11 Normal plot of residuals

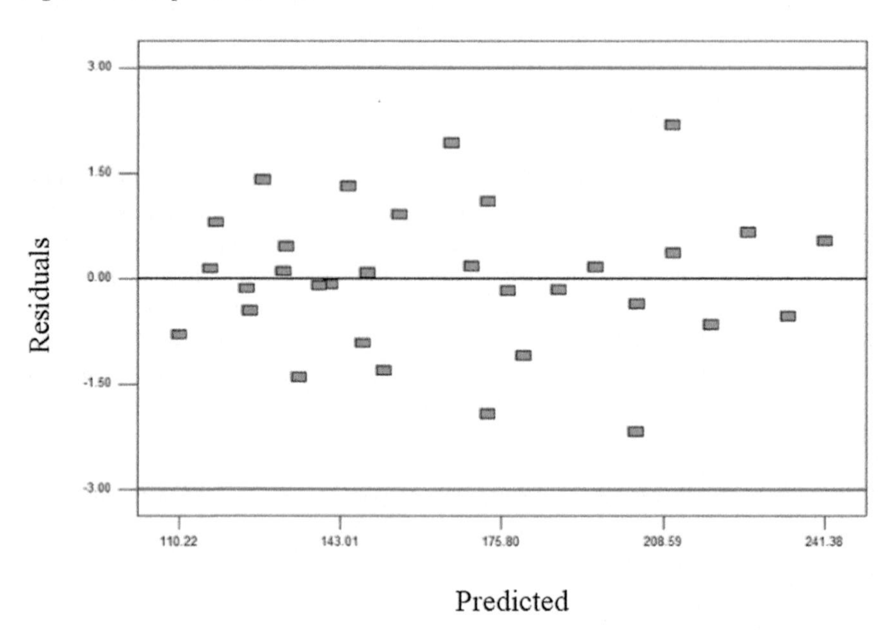

Fig. 12 Residual versus predicted value

Table 15 Optimum parameters from DOE

Heat generation					
Optimum parameters	Pin profile	Rotation speed (rpm)	Welding speed (mm/min)	Temperature (°C)	Desirability
	Threaded Cylindrical	1600	100	459	0.99
Peak stress					
Optimum parameters	Pin profile	Rotation speed (rpm)	Welding speed (mm/min)	Stress(MPa)	Desirability
	Threaded cylindrical	1600	100	237.68	0.94

4 Conclusion

Experimental analyses and DOE are performed by different rotational and welding speeds with various pin profiles. During experimental step 11 samples are approved for further studies due to absence of weld defect.

- In regard to heat generation, the results of experimental and DOE are almost similar.
- Increasing the rotational speed will increase temperature. On the other hand, by increasing the welding speed, temperature comes down.
- Highest strength produced with higher rotational speed due to better intermixing of the materials during the welding process.
- In aspect of microstructure, With the increase of rotational speed, the welding speed grain size would be bigger. Moreover, pin profiles have an influence on the microstructure and grain size after welding.
- The results of experiments and DOE for heat generation, tensile test and microstructure analysis indicate that sounds joint produced with threaded cylindrical tool pin profile with rotation speed of 1600 rpm and welding speed of 100 mm/min.

Acknowledgements The authors would like to acknowledge Universiti Teknologi PETRONAS, Malaysia for the financial support under I-Gen grant (0153DA-135).

References

1. Thomas W et al International patent no. 1991, PCT/GB92/02203, GB Patent
2. Meyghani B et al (2017) Developing a finite element model for thermal analysis of friction stir welding by calculating temperature dependent friction coefficient. In: 2nd International conference on mechanical, manufacturing and process plant engineering. Springer, Berlin

3. Mandal S, Rice J, Elmustafa A (2008) Experimental and numerical investigation of the plunge stage in friction stir welding. J Mater Process Technol 203(1):411–419
4. Meyghani B et al (2017) A comparison of different finite element methods in the thermal analysis of Friction Stir Welding (FSW). Metals 7(10):450
5. Threadgill PL, Leoneard AJ, Shercliff HR, Withers PJ (2009) Friction stir welding of aluminium alloys. [cited 2014; Available from: http://www.twi-global.com/technical-knowledge/published-papers/friction-stir-welding-of-aluminium-alloys/]
6. Meyghani B, Awang M, Emamian S (2017) A mathematical formulation for calculating temperature dependent friction coefficient values: application in Friction Stir Welding (FSW). In: Defect and diffusion forum. Trans Tech Publ
7. Mandal S, Williamson K (2006) A thermomechanical hot channel approach for friction stir welding. J Mater Process Technol 174(1–3):190–194
8. Lienert T et al (2003) Friction stir welding studies on mild steel. Welding J-NY 82(1):1-S
9. Hajideh MR et al (2017) Investigation on the effects of tool geometry on the microstructure and the mechanical properties of dissimilar friction stir welded polyethylene and polypropylene sheets. J Manuf Process 26:269–279
10. Meyghani B, Awang M, Emamian S (2016) A comparative study of finite element analysis for friction stir welding application. ARPN J Eng Appl Sci 11:12984–12989
11. Elangovan K, Balasubramanian V (2007) Influences of pin profile and rotational speed of the tool on the formation of friction stir processing zone in AA2219 aluminium alloy. Mater Sci Eng, A 459(1–2):7–18
12. Elangovana K, Elangovana V (2008) Influences of tool pin profile and welding speed on the formation of friction stir processing zone in AA2219 aluminium alloy. Mater Process Technol 200(1–3):163–175
13. Patil H, Soman S (2010) Experimental study on the effect of welding speed and tool pin profiles on AA6082-O aluminium friction stir welded butt joints. Int J Eng Sci Technol 2(5):268–275
14. Xu W et al (2013) Influence of welding parameters and tool pin profile on microstructure and mechanical properties along the thickness in a friction stir welded aluminum alloy. Mater Des 47:599–606
15. Thube RS, Pal SK (2014) Influences of tool pin profile and welding parameters on Friction stir weld formation and joint efficiency of AA5083 Joints produced by Friction Stir Welding. Magnesium (Mg) 4:4.9
16. Motalleb-nejad P et al (2014) Effect of tool pin profile on microstructure and mechanical properties of friction stir welded AZ31B magnesium alloy. Mater Des 59:221–226
17. Trimble D, O'Donnell GE, Monaghan J (2015) Characterisation of tool shape and rotational speed for increased speed during friction stir welding of AA2024-T3. J Manuf Process 17:141–150
18. Ilangovan M, Rajendra Boopathy S, Balasubramanian V (2015) Effect of tool pin profile on microstructure and tensile properties of friction stir welded dissimilar AA 6061–AA 5086 aluminium alloy joints. Defence Technol 11(2):174–184
19. Krasnowski K, Hamilton C, Dymek S (2015) Influence of the tool shape and weld configuration on microstructure and mechanical properties of the Al 6082 alloy FSW joints. Arch Civil Mech Eng 15(1):133–141
20. Nadikudi BKB et al (2015) Formability analysis of dissimilar tailor welded blanks welded with different tool pin profiles. Trans Nonferrous Metals Soc China 25(6):1787–1793
21. Emamikhah A et al (2014) Effect of tool pin profile on friction stir butt welding of high-zinc brass (CuZn40). The Int J of Adv Manuf Technol 71(1):81–90
22. Elangovan K, Balasubramanian K (2008) Influences of tool pin profile and tool shoulder diameter on the formation of friction stir processing zone in AA6061 aluminium alloy. Mater Des 29(2):362–373
23. Vijay SJ, Murugan N (2010) Influence of tool pin profile on the metallurgical and mechanical properties of friction stir welded Al–10 wt.% TiB2 metal matrix composite. Mater Des 31(7):3585–3589

24. Khodaverdizadeh H, Heidarzadeh A, Saeid T (2013) Effect of tool pin profile on microstructure and mechanical properties of friction stir welded pure copper joints. Mater Des 45:265–270
25. Mishra RS, Ma ZY (2005) Friction stir welding and processing. Mater Sci Eng R: Rep 50(1–2):1–78
26. Nandan R, DebRoy T, Bhadeshia HKDH (2008) Recent advances in friction-stir welding—process, weldment structure and properties. Prog Mater Sci 53(6):980–1023
27. He X, Gu F, Ball A (2014) A review of numerical analysis of friction stir welding. Prog Mater Sci 65:1–66
28. Emamian S et al (2017) A review of friction stir welding pin profile. Springer, Berlin
29. Awang Mucino VH, Feng Z, David SA (2005) Thermo-mechanical modeling of Friction Stir Spot Welding (FSSW) process: use of an explicit adaptive meshing scheme. SAE technical paper, 2006. 1:1251
30. Mandal S, Rice J, Elmustafa AA (2008) Experimental and numerical investigation of the plunge stage in friction stir welding. J Mater Process Technol 203(1–3):411–419
31. William EB (2005) Heat treatment, selection, and application of tool steels, in heat treatment, selection, and application of tool steels. Carl Hanser Verlag GmbH & Co. KG., pp I–XV
32. ASTM E3-01 (2001) Standard practice for preparation of metallographic specimens. ASTM International, West Conshohocken
33. ASTM E407-07 (2015) Standard practice for microetching metals and alloys. ASTM International, West Conshohocken
34. Zahraee SM et al (2013) Combined use of design of experiment and computer simulation for resources level determination in concrete pouring process. Jurnal Teknologi 64(1):43–49
35. Sadeghifam AN et al (2015) Combined use of design of experiment and dynamic building simulation in assessment of energy efficiency in tropical residential buildings. Energy Build 86:525–533
36. Chen C, Kovacevic R (2003) Finite element modeling of friction stir welding—thermal and thermomechanical analysis. Int J Mach Tools Manuf 43(13):1319–1326
37. Zhang H, Zhang Z, Chen J (2005) The finite element simulation of the friction stir welding process. Mater Sci Eng, A 403(1):340–348
38. Rodriguez R et al (2015) Microstructure and mechanical properties of dissimilar friction stir welding of 6061-to-7050 aluminum alloys. Mater Des 83:60–65
39. Montgomery D (2009) Basic experiment design for process improvement statistical quality control. Wiley, USA
40. Zahraee SM et al (2014) Application of design of experiment and computer simulation to improve the color industry productivity: case study. Jurnal Teknologi 68(4):7–11
41. Zahraee SM et al (2014) Application of design experiments to evaluate the effectiveness of climate factors on energy saving in green residential buildings. Jurnal Teknologi (Sci Eng) 69(5):107–111

The Effect of Argon Shielding Gas Flow Rate on Welded 22MnB5 Boron Steel Using Low Power Fiber Laser

Khairul Ihsan Yaakob, Mahadzir Ishak and Siti Rabiatull Aisha Idris

Abstract This study deals with an investigation of shielding gas flow rate on continuous wave (CW) and pulse wave (PW) mode of welded boron steel (22MnB5) using low power fiber laser. Argon gas is selected as shielding gas. The observation of welding surface, geometry, microstructure and hardness distribution were carried out with different shielding gas flow rate from 5 to 25 L/min. The result found 15 L/min is the optimum argon shielding gas flow rate to produce good weld surface and deeper penetration which is apparent in PW mode application. The microstructure and mechanical properties are not affected by shielding gas flow rate. It is majorly influenced by thermal experience during particular welding process.

Keywords Welding · Laser processing · Boron Steel · Light weight

1 Introduction

Boron steel (22MnB5) is one type of Advance High Strength Steel (AHSS) replacing the usage of high-strength steel (HSS) material used in car body [1, 2]. The joining method for automotive component shows the interest in laser welding due to its deep penetration, high speed, and small heat affected zone especially by using fiber laser [3–10].

In laser welding process, the shielding gas is commonly fed into the weld area by side nozzle which is responsible for protecting the weld pool against weld atmosphere, protecting the optics from spatter and also suppressing the plasma plume [11]. The plasma plume by the mix of metal vapor from keyhole and shielding gas composed of ionized gas, molten metals, slags, vapors and gaseous atom. In 2010 [1], plasma plume was found could influence the processing condition through absorption of the laser radiation and change the energy transfer from the laser beam to the

K. I. Yaakob · M. Ishak (✉) · S. R. A. Idris
Faculty of Mechanical Engineering, Universiti Malaysia Pahang, 26600 Pekan, Pahang, Malaysia
e-mail: mahadzir@ump.edu.my

M. Awang (ed.), *The Advances in Joining Technology*, Lecture Notes in Mechanical Engineering, https://doi.org/10.1007/978-981-10-9041-7_3

material. In 2011 [2], reported that the Helium have the best ability in suppressing the plasma formation for CO_2 laser welding after comparing with several types of gasses. He also stated that Argon would produce larger plasma plume due to its heavy properties and low ionization potential depending on laser power. Those findings were based on the CO_2 laser which was highly affected by the plasma plume formation; different from the solid state laser usage which has a smaller wavelength.

In the previous finding [3], it stated that the plasma cloud is almost transparent to a small wavelength of lasers such as Nd:YAG and Fiber laser. However, the transparency of plasma cloud with a small wavelength laser will reduce by decreasing the welding speed due to insufficient to leave the plasma behind. Another plasma plume optimization by using Nd:YAG laser was done by [12]. They also reported that the ionization potential of process gas has no obvious importance in plasma plume control. The result of their study about the flow rate of shielding gas to remove plasma plume depends on the gas properties such as weight and also shielding gas setup parameters such as the nozzle size and positioning of the jet. Supported by the latest finding [13], there is no effect of blowing distance. Moreover, the plasma size decreased as the gas flow rate increased without affecting the stability of the keyhole. Even though the small wavelength of laser type gives transparency of plasma cloud, welding speed and gas flow rate is crucial in producing a high-quality welding; particularly in the application of low power laser which requires slower welding speed.

So far, the investigation on the effect of shielding gas and optimization process has been conducted using galvanized steel for automotive application. Yet, very few studies have examined on the specific type of steel such as boron steel using low power laser. Therefore, the objective of this present work is to determine the effect of shielding gas flow rate on welded boron steel (22MnB5) using low power fiber laser with continuous wave (CW) and pulse wave (PW) welding mode. The bead on plate (BOP) welding is conducted using the various flow rate of Argon gas for both welding mode. The observation on welding surface, penetration depth, microstructure and hardness distribution is carried out.

2 Experimental Setup

1.6 mm thick of 22MnB5 was used in this study. The material composition of this boron steel is shown in Table 1. Al–Si coating on the as-received material surface was removed by using the surface grinding machine. The specimens were cut to 50 mm × 50 mm dimension sufficient BOP welding.

Table 1 Material composition

Composition (%)	C	Si	Mn	P	S	Cr	Ni	B
Boron steel	0.26	0.30	1.24	0.016	0.003	0.20	0.016	0.004

Fig. 1 Experiment setup

The IPG-YLM-QCW series of a fiber laser with an average power of 200 W was used for CW and PW mode. Laser power and welding speed were fixed at 50% from average power and 5 mm/s respectively for both welding mode. 1 ms of pulse width and 60 Hz of pulse repetition rate were fixed for PW welding. The focal length of this fiber laser is 200 mm. The focus point was placed on the top of the specimen's surface. The incident angle was set at 5° to avoid reflection of the beam. The diameter of shielding gas nozzle was 6 mm, and the nozzle tip distance to the beam was 30 mm with 20° of angle. The experiment setup is shown as in Fig. 1. Argon was used as the shielding gas, and its flow rate varied from 5 to 25 L/min as a variable process parameter.

The welded surface observation was carried out using optical macroscope. For depth, width and microstructure observation, optical microscope was used on the prepared samples. The welded cross section samples were finely cut and mounted. Then, the samples were grinded, polished and etched using 2% of Nital solution to reveal the microstructure. The width and depth from the observation of welding cross section were taken based on the microstructure changes. By using similarly mounted samples, the hardness test was conducted on optimum gas flow rate sample at 50% of penetration depth across the welded area using 0.5 HV.

3 Results and Discussion

3.1 Surface Observation

Figures 2 and 3 show the surface of specimens with CW and PW welding with different shielding gas flow rate. It was observed that the welding surface of 5 L/min for both welding modes show the burn mark with irregular weld width surface appearance. The higher flow rate for CW welding shows the increasing of slag deposition

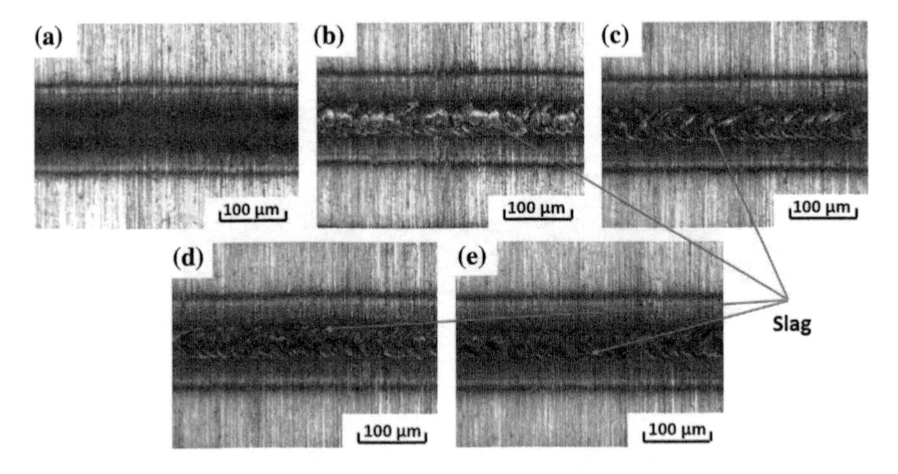

Fig. 2 CW welding surface **a** 5 L/min, **b** 10 L/min, **c** 15 L/min, **d** 20 L/min, **e** 25 L/min

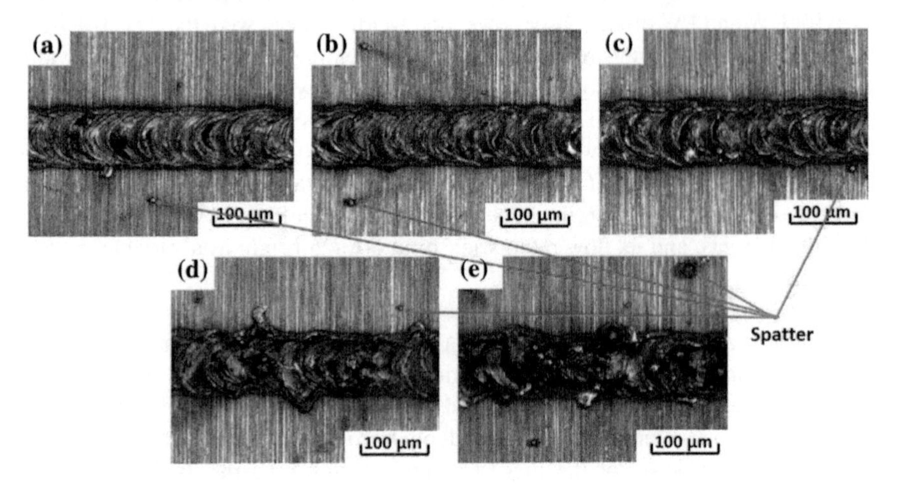

Fig. 3 PW welding surface **a** 5 L/min, **b** 10 L/min, **c** 15 L/min, **d** 20 L/min, **e** 25 L/min

on the weld surface. Unlike the PW welding where the slag deposition is almost uniform, when the gas flow rate is increased.

The slag existence on the weld surface was due to an involvement of the plasma plume. The plasma plume is composed of ionized gas, molten metals, slags, vapors and gaseous atoms and molecules [14]. In this case, the slag from plasma plume was deposited on the weld surface during the solidification process. The increasing of gas flow rate will increase the volume of plasma plume over the welding surface causing the amount of deposited slag to increase. However, the difference of slag formation between CW and PW mode of weld might be due to the fluid flow and surface tension on the molten pool shown in Fig. 4 [15].

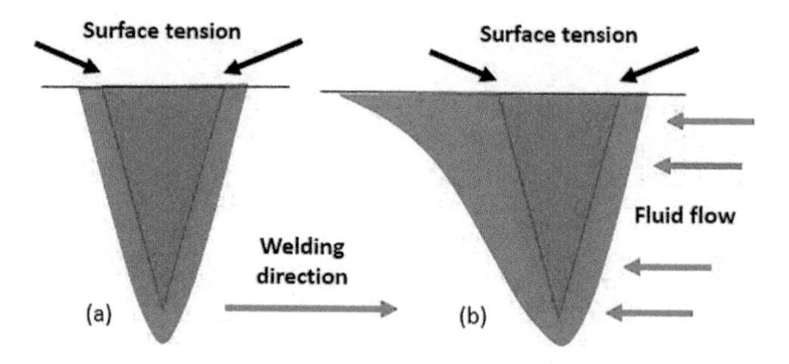

Fig. 4 Schematic for **a** PW, **b** CW welding

PW welding is stationary weld and only faces the surface tension gradient caused by a temperature gradient. It was supported by the fact that the weld pool freezes between pulses. This effect will produce the constant formation of slag on the welding surface. CW welding was also experiencing the same effect with additional weld pool movement effect. The movements of molten pool drag the slag inside plasma plume along the welding line. The increment of slag with higher gas flow rate lead to an irregular amount of slag deposited on the molten pool surface, and it can be apparently observed. However, the slag does not exist inside the molten pool and might not affect the strength and properties of welding due to slag impurities.

Pertaining to the recent work, the increment of the shielding gas flow rate increased the pressure heading to the molten pool and created the chaos formation lead to the non-uniform weld bead produced. This result can be observed in PW mode welding with the non-uniform overlapped pulses at 20 and 25 L/min. The CW mode welding was also affected by this condition where the turbulent formation of the molten pool might blend the slag together and caused it to be deposited on the weld surface.

Pertaining to the recent work, the increment of the shielding gas flow rate increased the pressure heading to the molten pool and created the chaos formation lead to the non-uniform weld bead produced. This result can be observed in PW mode welding with the non-uniform overlapped pulses at 20 and 25 L/min. The CW mode welding was also affected by this condition where the turbulent formation of the molten pool might blend the slag together and caused it to be deposited on the weld surface.

3.2 *Width and Depth Observation*

Figure 5 shows the cross section of BOP welding of PW and CW mode. The penetration of PW mode welding was higher compared to CW mode welding due to the difference in peak power density. The values of peak power density for PW and CW mode were calculated as 1.515 and 0.113 MW/cm^2 respectively. In another point of view, the interaction time is the heating time of the process on the weld centerline which affects the penetration. In this study, interaction time of CW mode was higher

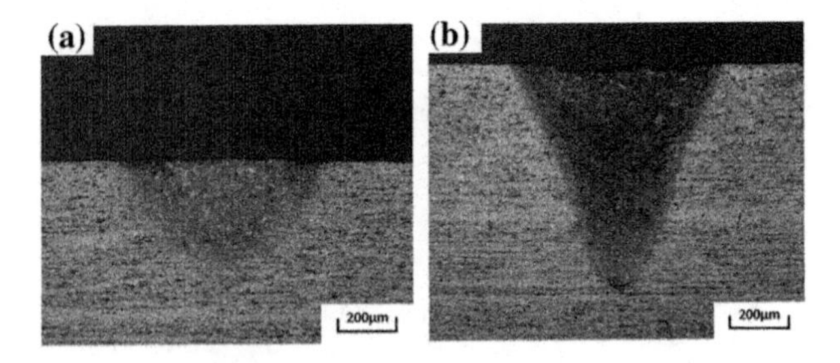

Fig. 5 Welding cross section of 15 L/min Argon flow rate **a** CW, **b** PW

Fig. 6 Depth of penetration versus flow rate **a** CW, **b** PW

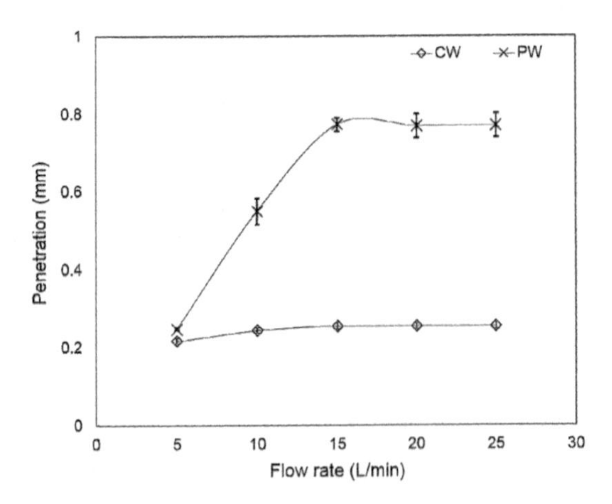

compared to PW mode. Contradictory to the research from [15] which stresses on the effect of interaction time between these two modes, the peak power density shows significant influences in penetration compared to interaction time. However, high interaction time in CW mode shows wider HAZ compared to shorter interaction time of PW mode.

For the penetration observation, Fig. 6 shows the penetration depth of the CW and PW mode welding with the increasing gas flow rate. The penetration depth between CW and PW mode welding shows a similar pattern with a different value. The increment of penetration starts at 5 L/min and stops at 15 L/min. The penetration remains constant at 15 L/min and above.

The penetration pattern can be explained by the findings from the previous research. The existence of plasma plume over the workpiece will reduce the penetration because the laser energy will be absorbed by the plasma plume and subsequently, reducing the energy reach on the workpiece surface. The balance of energy will hit the surface and create small penetration. However, the suitable amount of plasma plume inside the molten pool is helpful by enhancing the absorption of laser energy to the workpiece [4]. The Fiber laser has a small wavelength compared to CO_2 laser. The smaller laser wavelength makes it easier to penetrate through plasma over keyhole and reduces the influence of the plasma plume. However, with lower speed, reported below 17 mm/s, the laser beam is needed to encounter the plasma cloud, and it will affect the energy reach to the surface and create a shallow penetration [16]. The ionization potential of the process gas has no obvious importance in fiber laser welding. This observation differs from what is commonly known for the high power of CO_2 laser welding (at travel speeds of less than 17 mm/s) where Helium is, by far, the best process gas for plasma control due to its high ionization potential and thermal conductivity [17]. Thus, the most influenced factor that affected the penetration on the recent study was the shielding gas flow rate.

From the results of the present study, the 5 L/min of argon flow rate was insufficient to remove the plasma cloud over the surface causing the small penetration. With the increasing gas flow rate, the volume of plasma cloud will reduce and also increase he penetration. The maximum penetration shows the optimum volume of plasma plume needed for the molten pool to enhance the penetration. The penetration achieved the maximum value starting from 15 L/min and remained constant with an increasing gas flow rate until 25 L/min. However, the high flow rate of shielding gas gave pressure to the molten pool and created the unstable formation and bad weld surface. Thus, the stability of molten pool needs to be parallel with the consistency of plasma plume over the molten pool to enhance the penetration [13].

3.3 Microstructure and Mechanical Properties

The microstructure observation at fusion zone (FZ) and heat affected zone (HAZ) remain the same with increasing shielding gas flow rate. However, the difference of microstructure exists when comparing between CW and PW mode. Figure 7 shows the microstructure of 15 L/min of CW mode and PW mode from FZ area across HAZ. At FZ, both specimens show mainly martensite microstructure. Additionally, CW mode specimen consisted of wider HAZ compared to PW mode and exposed the coarse grain HAZ (CGHAZ).

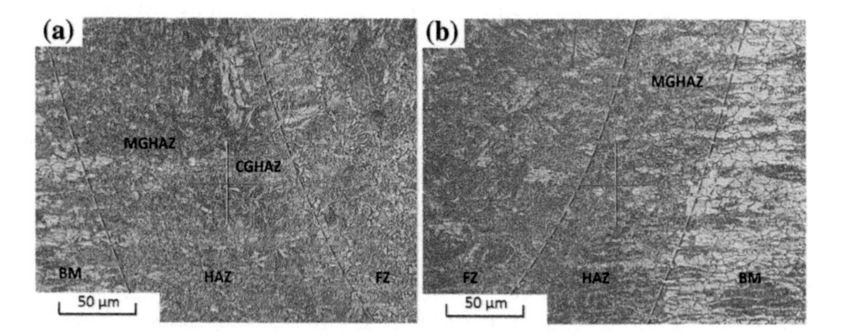

Fig. 7 microstructure of 15 L/min samples for **a** CW, **b** PW mode across HAZ area

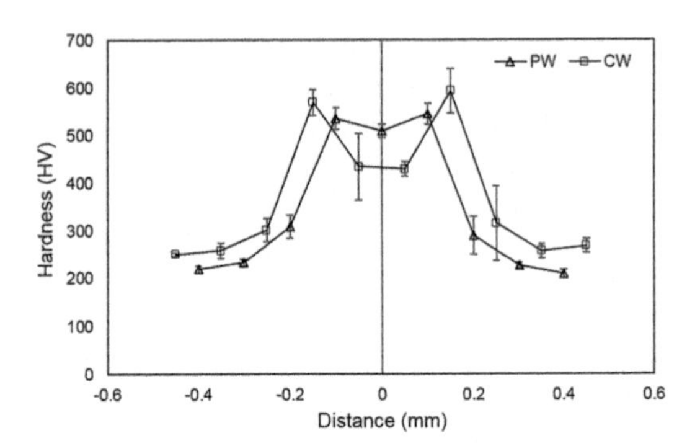

Fig. 8 Hardness distributions across weld area for CW and PW mode

The mechanical properties of welded area are evaluated on hardness test. Figure 8 shows hardness pattern across the weld area for 15 L/min of the gas flow rate of CW and PW mode. The hardness profile shows a reduction of hardness value at the center of the weld. Wider weld area and huge hardness reduction is detected for CW mode compared to PW mode. These hardness properties might vary due to the microstructure transformation according to heat input and cooling rate during the welding process.

The hardenability of boron steel is resulting from boron addition and strongly related to the segregation behavior of boron which suppresses the ferrite segregation. Non-equilibrium segregation is closely related to welding application due to cooling rate influence. The segregation of boron initially increased and then decreased with an increase in the heat input. It is due to the back-diffusion of boron as a result of an increase in the exposure time at high temperature after non-equilibrium grain boundary segregation. Moreover, its hardness also reduces with increase heat input and slower cooling rate [18–21].

Relating to recent work, both FZ experienced peak temperature above the liquidus temperature. The high heat input is responsible for hardness reduction in this area. However, the hardness of CW mode welding is lower than PW mode. During CW mode welding, FZ experience slower cooling rate compared to PW mode. It is because molten pool of CW mode is dragged along weld area which delays the solidification process and also reduces its cooling rate. This is in contrast with PW mode welding which solidified faster in each pulse. Thus, the hardness degradation is less compared to CW mode.

Meanwhile, at HAZ, CW mode produces wider area with CGHAZ existence which mainly consists of hard lath martensite microstructure. It is due to high heat input from FZ spread to this area and sufficient for austenite transformation. Thus, it leads to higher cooling rate experienced in this area. Contrary to the small area of mix grained HAZ (MGHAZ) which might consist of bainite at PW mode, the heat input at FZ experienced very high cooling rate and the temperature spread to HAZ area is insufficient for austenite transformation. Thus, the hardness of HAZ area for PW mode is lower than CW mode.

4 Conclusions

The purpose of current study is to determine the effect of shielding gas flow rate on boron steel by using a low power fiber laser. The effect was determined by using Argon gas on the welding surface and penetration depth including microstructure and mechanical properties observation.

1. Plasma plume occurs at PW and CW mode. However, the effect is apparently observed on welding mode with deep penetration.
2. Plasma plume phenomenon can be resolved by applying 15 L/min of argon gas flow rate according to a sound characteristic of penetration depth and weld surface.
3. Shielding gas flow rate does not affect the microstructure and hardness properties of welded cross section. Thermal experience between PW and CW mode are significant influencers in both characteristics.

Acknowledgements This document was prepared to help authors. The authors are grateful to the Universiti Malaysia Pahang for providing financial support given under research grant RDU180314.

References

1. Gan W, Babu SS, Kapustka N et al (2006) Microstructural effects on the springback of advanced high-strength steel. Metall Mater Trans A 37:3221–3231
2. Xiaodong Z, Zhaohui M, Li W (2011) Current status of advanced high strength steel for auto-making and its development in baosteel (Shanghai)
3. Suder WJ, Williams S (2014) Power factor model for selection of welding parameters in CW laser welding. Opt Laser Technol 56:223–229
4. Mei L, Chen G, Jin X et al (2009) Research on laser welding of high-strength galvanized automobile steel sheets. Opt Lasers Eng 47:1117–1124
5. Bardelcik A, Worswick MJ, Wells MA (2014) The influence of martensite, bainite and ferrite on the as-quenched constitutive response of simultaneously quenched and deformed boron steel—experiments and model. Mater Des 55:509–25
6. Moskvitin GV, Polyakov AN, Birger EM (2013) Application of laser welding methods in industrial production. Weld Int 27:572–80
7. Kawahito Y, Mizutani M, Katayama S (2007) Investigation of high-power fiber laser welding phenomena of stainless steel. Trans JWRI 36:11–16
8. Hetch J (2012) Fiber lasers: the state of the art. Laser Focus World, 1–22
9. Mei L, Yan D, Yi J et al (2013) Comparative analysis on overlap welding properties of fiber laser and CO_2 laser for body-in-white sheets. Mater Des 49:905–912
10. Assunção E, Quintino L, Miranda R (2009) Comparative study of laser welding in tailor blanks for the automotive industry. Int J Adv Manuf Technol 49:123–131
11. Reisgen U, Schleser M, Mokrov O et al (2010) Shielding gas influences on laser weldability of tailored blanks of advanced automotive steels. Appl Surf Sci 257:1401–1406
12. Gerritsen CHJ, Olivier CA (2002) Optimization of plasma/plume control for high power Nd:YAG laser welding of 15 mm thickness C–Mn steels. In: 6th international conference in welding research
13. Li K, Lu F, Cui H et al (2014) Investigation on the effects of shielding gas on porosity in fiber laser welding of T-joint steels. Int J Adv Manuf Technol, 1881–1888

14. Kah P, Martikainen J (2013) Influence of shielding gases in the welding of metals. Int J Adv Manuf Technol 64:1411–1421
15. Assuncao E, Williams S (2013) Comparison of continuous wave and pulsed wave laser welding effects. Opt Lasers Eng 51:674–680
16. Fisher S, Olivier CA and Riches ST (1999) Optimisation of plasma control parameters for Nd:YAG laser welding of stainless steel enclosures. In: Nordic conference in laser processing of materials
17. Chung B, Rhee S, Lee C (1999) The effect of shielding gas types on CO_2 laser tailored blank weldability of low carbon automotive galvanized steel. Mater Sci Eng A 272:357–362
18. Kim S, Kang Y, Lee C (2016) Effect of thermal and thermo-mechanical cycling on the boron segregation behavior in the coarse-grained heat-affected zone of low-alloy steel. Mater Charact 116:65–75
19. Yang H, Wang X, Qu J (2014) Effect of boron on CGHAZ microstructure and toughness of high strength low alloy steels. J Iron Steel Res Int 21:787–792
20. Wang XN, Chen CJ, Wang HS et al (2015) Microstructure formation and precipitation in laser welding of microalloyed C–Mn steel. J Mater Process Technol 226:106–114
21. Kim S, Kang Y, Lee C (2013) Variation in microstructures and mechanical properties in the coarse-grained heat-affected zone of low-alloy steel with boron content. Mater Sci Eng A 559:178–186

Effect of Bevel Angle and Welding Current on T-Joint Using Gas Metal Arc Welding (GMAW)

Z. A. Zakaria, M. A. H. Mohd Jasri, Amirrudin Yaacob, K. N. M. Hasan and A. R. Othman

Abstract In this study, the effects of various welding parameters on welding strength in mild steel A36, welded by gas metal arc welding with fillet joint under 1F position were investigated. The welding current and the angle of bevel are the variable parameters, while welding speed was chosen as constant parameter. Each specimen's mechanical properties have been measured after the welding process, and the effects of the welding parameters on the strength were investigated. Then, the relationship between welding parameter and mechanical properties is determined. The project used 8 specimens to be studied and to find the best welding parameters for mild steel plate on fillet joint (1F position). Based on the mechanical tests performed, the best welding parameters for the mild steel plate thickness of 5 mm T-joint fillet welds are obtained; it was 120 A for weld current with no bevel angle.

Keywords Welding · Welding parameter · Tensile · Hardness · Microstructure

Z. A. Zakaria (✉) · M. A. H. Mohd Jasri · A. Yaacob
Universiti Kuala Lumpur, Malaysian Institute of Marine Engineering Technology, Lumut, Perak, Malaysia
e-mail: zainul@unikl.edu.my

M. A. H. Mohd Jasri
e-mail: azrie@unikl.edu.my

A. Yaacob
e-mail: amirrudin@unikl.edu.my

K. N. M. Hasan · A. R. Othman
Universiti Teknologi PETRONAS, Seri Iskandar, Perak, Malaysia
e-mail: khairulnisak.hasan@petronas.com.my

A. R. Othman
e-mail: rahim.othman@petronas.com.my

M. Awang (ed.), *The Advances in Joining Technology*, Lecture Notes in Mechanical Engineering, https://doi.org/10.1007/978-981-10-9041-7_4

1 Introduction

Gas metal arc welding (GMAW) is one of the Arc Welding processes that is widely used in the industries nowadays. GMAW is a process where the continuous fed of the wire electrode or filler wire through the welding gun and the shielding gas on the base metal and there will be no slag covering on created welding bead. GMAW process was introduced to the industry in the late of 1940's by Hobart and Devers. Aluminium wire was the first electrode used and argon gas as the shielded gas. In the early 1950's the first GMAW process was developed for the steel metal by using steel electrode by Lyubavshkii and Novoshilov, but due to high weld spatter, this process was not very user friendly. As time passes, this process has been through lot of innovate and now become one of the important welding process [1].

Butt, corner, tee, lap, and edge are the basic types of welding joint. The right angle joint of two members are called tee joint. In cross section, letter "T" shape appears for t joint and L-shape for the corner joints. This type of joint is under fillet type of welding. The joining of two surfaces materials with the right angle shape in triangle to each other is under fillet weld type. The effect of bevel angles and welding current on the mechanical properties of fillet joint under 1F position was studied in this research.

2 Literature Review

2.1 Welding Parameters

The proper choice of weld parameters can result in a high quality of weld. However, these variables cannot be changed by single parameters only, and to achieve the desired result, others parameters need also to be changed rather than changing one variable only. A good set up of these variables or in the other word the best recipe of the setting will result in high quality weld metal [2].

2.2 The Effect of Welding Current

The correct setting of current will not disappoint the end result. Too low current setting may cause insufficient heat to melt the metal base. Thus it will make the molten pool to be too small, pile up, and look irregular. However, if the current setting is too high, it will cause the weld molten pool to be large and irregular due to overheating and the fillet wire melting too fast [3].

2.3 *The Effect of Bevel Angle*

It is discovered that better mechanical properties can result from the preparation of the V-groove edge. Based on this case study, it is identified that more amount of weldment deposit on material with V-groove edge compared to straight edge [4]. It was suggested that smaller angle is better as long as the angle of groove can complete a penetration groove [5].

3 Experimental Works

The material for the plate is mild steel or low carbon steel is ASTM A36. This type of metal has high welding properties and it is eligible for punching, drilling grinding and machining process. It is a common steel grade being used for structure. A36 has high yield strength, thus making it have a high bending capability [6].

The specimens that are used to carry out this project are ASTM A36 low carbon steel with the dimension of 70 mm × 300 mm × 5 mm flat bar as in Fig. 1. In this project the joint type is t-joint and welded under 1F position with one sequence layer on both side. 45° angle of bevel was selected to the variable in this project. The welding machine that is used to perform this welding is PHOENIX 330 Expert Pulse force Arc. This task or project used GMAW welding process manually. The welding process parameters distribution is shown as in Table 1. Three mechanical tests were executed on the welded plate, tensile test, Brinell Hardness test and macro test.

Fig. 1 Welding test sample

4 Testing

4.1 Tensile Test

Tension test is another word for the tensile test. It is a common type of mechanical tests performed on material. When the material being pulled, the strength will be measured by how much it will elongate. Then, the specimens were assembled with a clamp or jig. The bolt and nut need to be used to ensure the clamp and the specimen was strongly attached as in Fig. 2. Tensile test machine by Victor brand was used to complete this test. The total specimens for tensile test in this project were 27 pieces for each panel. All of these were done with the additional clamp or jig with the speed flow rate of 1 mm/min.

4.2 Brinell Hardness Test

Brinell hardness testing is usually done on material that has structure and rough surface. Typically for iron and steel material, the applied force is up to 3000 kg with

Table 1 Welding process parameters

Parameters	No.	Current (A)	No.	Angle (°)
Variables values	1	90	1	0
	2	100		
	3	110	2	45
	4	120		

Fig. 2 Specimen was set at the machine by using flat face wedges grip and firmly holds

10 mm diameter carbide ball as the intender. Specimens that are used for hardness test are 16 specimens. 187.5 kgf was applied on the intender ball with 2.5 mm diameter.

5 Result and Discussion

5.1 Ultimate Tensile Strength

In these experiments, it used 8 specimen with different welding parameters. From Fig. 3, the tensile results obtained show that maximum tensile strength of 377 MPa is possessed by the specimens made using 120 A weld current and non-bevel. The medium value of the tensile strength was 309 MPa, shown by the specimen with the current value 100 A and non-bevel. The lowest value of ultimate tensile strength was the specimen with the setting parameters of 90 A current and 45° angle of bevel that is 272 MPa.

As stated above, welding current of 120 A welding current and no bevel angle has a high tensile strength and ductility. Considering that, this is the right setting of welding parameters for GMAW process. Higher the voltage and ampere will give the effect on the heat affected zone of the weld area, especially during the solidification process between the filler and base metal.

In this project, bevel angle does not show lot of differences in UTS result and this may be due to thickness of plate that too thin which is 5 mm. The high value of UTS in this project was on no bevel angle specimen and the lowest value of UTS is on the specimen with bevel angle. Young modulus for tensile test was calculated and analysed after the test and the result can be seen in Fig. 4.

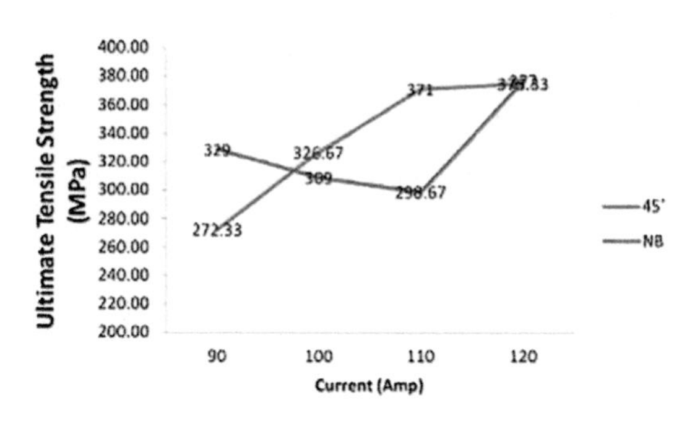

Fig. 3 Ultimate tensile strength (MPa) vs welding current

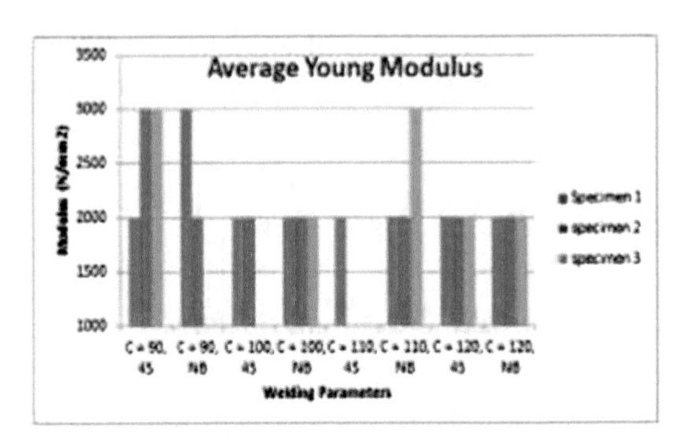

Fig. 4 Young Modulus versus welding current

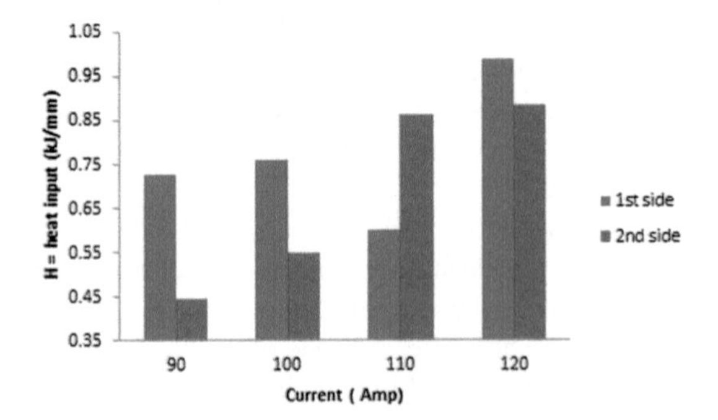

Fig. 5 Heat input versus current for 45° bevel angle

5.2 *Average Young Modulus*

The welding current of 90 A and with 45° bevel angle specimen shows the highest modulus elastic result. This shows that the specimen can stand with the maximum force imposed which means that it is not brittle or elastic. The lowest modules elastic result is shown by specimen of 110 A current value and 45° bevel angle. The average value of modules elastic was shown by specimen with welding parameters 120 A current. From this result, identified that specimen with 120 A current value is the best because its average is not brittle and elastic at the same time not too strong because it will cause wasted sources.

5.3 *Heat Input*

Based on this experiment, every specimen has 2 sides of weld. This section will shows the example of heat input calculation. Thus the data of the heat input in this project was tabulated in table and shown in Figs. 5 and 6. The formula of heat input as shown below.

$$H = I \times V \times 60/\,(S \times 1000)$$

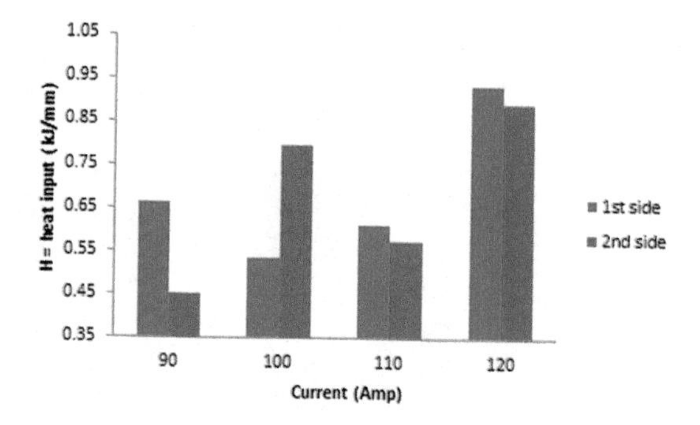

Fig. 6 Heat input versus current for no bevel angle

where H is heat input (kJ/mm), I is current (A), V is arc voltage (V) and S is travel speed (mm/s).

5.4 *Brinell Hardness*

Two point indents have been made for each specimen (HAZ and base metal). The radius of the indention is then measured through a special microscope and data is recorded. Data obtained must be multiplied by the scale of 0.05 set in a special microscope. Then the formula of Brinell hardness number was applied. Formula of Hardness Brinell is shown below:

$$BHN = \frac{2F}{\pi D\left[D - \sqrt{(D^2 - d^2)}\right]}$$

In above equation, F is the applied force (kgf), D is diameter of indenter (mm) and d is diameter of indentation (mm).

Figure 7 showed the hardness results for the specimens with different welding currents. From the result, the hardness of the HAZ did not grow until it reached the base material. The highest value of BHN for base metal was 187 that by specimens with welding current 90, 100 and 110 A with angle of bevel was 45° and no bevel. The highest value of BHN for HAZ area was 170 A shown by specimens with welding parameters 100 A at no bevel, 110 A at 45° bevel angle and 120 A both with no angle and 45° bevel angle. The result data showed that 120 A was the highest value of BHN on the HAZ area. High BHN values show that the indention diameter is small. This means that, the strength of the specimen or material surface is great.

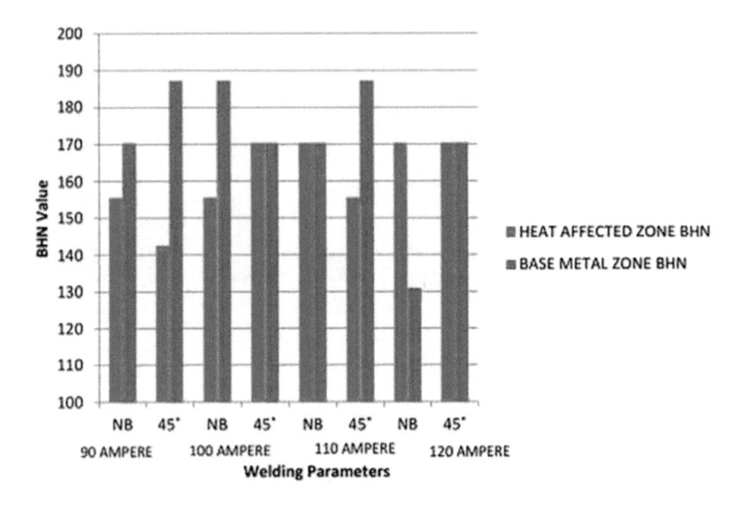

Fig. 7 Hardness Brinell number versus welding current

5.5 *Discussion*

From this experiment result of tensile test shows that, 377 MPa is possessed by the specimens made using 120 A weld current and non-bevel. The medium value of the tensile strength was 309 MPa that is the specimen with the current value 100 A and non-bevel. The lowest value of ultimate tensile strength was the specimen with the setting parameters of 90 A weld current and 45° angle of bevel that is 272 MPa. Hardness results for the weld shows that the highest value of BHN for base metal was 187 BHN shown by the specimens with welding current 90, 100 and 110 A with angle of bevel was 45° and no bevel. The highest value of BHN for HAZ area was 170 BHN by specimens with welding parameters 100 A no bevel, 110 A 45° bevel angle and 120 A both with no angle and 45° bevel angle.

The result of the heat input calculation as shown in Figs. 5 and 6 identified that, the highest value was specimen with welding parameters of 120 A current and no bevel angle that is 0.93 and 0.89 kJ/mm on both sides respectively. In relation to the tensile result, this specimen has a high value of UTS and during the tensile test, the specimen broke at the parent metal not at weldment area.

It can be concluded that the welding parameters for specimen 8 can be considered as the suitable and right set for the welding process. The current should be in high level because medium and lower level will settings affect the strength of the joint and may produce incomplete fusion due to not enough heat. No Bevel angle for t-joint of 5 mm thickness of mild steel is suggested based on the result found above.

6 Conclusions

As the conclusion, the objective of this experiment was achieved. The strength analysis on 1F welding position (fillet weld) on variable welding parameters by using tensile test and hardness has been done. The study determined that the process parameters to obtain maximum values for the weld properties were 120 A for welding current, and no bevel angle for low carbon steel plate of 5 mm thickness.

To choose between 90 and 120 A for the optimum parameter, it is essential to remember that the depth of penetration can give an extra point of successive welding. The current 120 A based on this experiment is optimum parameter than the others or below than that. Finally, the results of the mechanical test showed that, welding parameters influence the weld joint structure.

7 Recommendations

It is recommended that the research can be continued with other materials, plate thickness or welding process which can give better welding quality. Next researchers are encouraged to continue this experiment with different welding parameters, for example the voltage, heat input and etc. Pertaining to T-joint weld, the design of the clamp or jig for tensile test are recommended to have good preparation. The top part of the jig should stronger than the bottom part and the slot size of the welded specimen must be not too large to obtain accurate result of tensile test. The result obtained from the research experiment can be compared with the result obtained from this experiment. Based on the test result, this research can be further enhanced by increasing the number of testing specimens. So, the best of parameter and results obtained more accurate. Welding technology is a very large syllabus and a lot of things can be studied and investigated to produce high quality welding.

References

1. Conrardy C (2010) Gas metal arc welding, p. 2
2. Sivasakthivel K, Rajkumar R, Yathavan S (2015) Optimization of welding parameter in MIG welding by Taguchi method. In: Proceedings of international conference on advances in materials, manufacturing and applications, 761 (Amma), pp 761–765
3. Silva J, Silva B, De Brito MD, De Sousa R (2013) 5 essentials for good welding, 2
4. Bodude MA, Momohjimoh I (2015) Studies on effects of welding parameters on the mechanical properties of welded low-carbon steel, (May), 142–153
5. Chen J, Schwenk C, Wu CS, Rethmeier M (2012) Predicting the influence of groove angle on heat transfer and fluid flow for new gas metal arc welding processes. Int J Heat Mass Transfer 55(1–3):102–111. https://doi.org/10.1016/j.ijheatmasstransfer.2011.08.046
6. Beoh CHS, Giacomo S (1995) Chemical composition, 4

Laser Brazing Between Sapphire and Inconel 600

Shamini Janasekaran, Farazila Yusof and Mohd Hamdi Abdul Shukor

Abstract Joining ceramic to metal has become essential in many industries as the usage increased intensely. In this feasibility study, 99.999% artificial sapphire; a type of ceramic consists of Al_2O_3 was bonded using low power Ytterbium fiber laser to Inconel 600. An active filler alloy (Cusil ABA) was used in between the workpiece to achieve the adhesion effect on sapphire. Ranges of welding speed, laser power and focal distance were tested on the samples and the optimum parameter was obtained from the appearance on the sapphire surface. The bonding areas were analysed under scanning electron microscope and energy dispersive X-ray to view the braze ability of bonding region. The X-ray diffraction results determined the reactions between Inconel 600, active filler alloy and sapphire. The joining area formed "ocean" structures near the sapphire and Inconel that enables an adhesion layers to keep the workpiece together. Interface composition had revealed intermetallic layers formed on sapphire consisting of $AlNi_3$, Cr_2Ti and Fe_2Ti. Meanwhile, Inconel 600 had reacted further with Cusil ABA during laser brazing forming additional intermetallic layers of $AlCr_2$ and Al_3Ni_2.

1 Introduction

Brazing is one of the techniques used to join metals to ceramics [1]. The ceramic has superior properties such as high temperature stability, corrosion and wear resistance and these have become the main reasons for the industries to use ceramic in many field applications [2, 3]. In recent years, products with metal-ceramic bonding have

---

S. Janasekaran (✉)
Department of Mechanical Engineering, Faculty of Engineering and Built Environment, SEGi University, Kota Damansara, Petaling Jaya, Selangor, Malaysia
e-mail: shaminijanasekaran@segi.edu.my

F. Yusof · M. H. A. Shukor
Department of Mechanical Engineering, Faculty of Engineering, University of Malaya, Kuala Lumpur, Malaysia

M. Awang (ed.), *The Advances in Joining Technology*, Lecture Notes in Mechanical Engineering, https://doi.org/10.1007/978-981-10-9041-7_5

become a norm due to the excellent mechanical properties that leads to recommendation in aerospace, automobile and electronics fields. Joining metal and ceramic allows both properties of the dissimilar materials to be engaged in customized products [4, 5]. However, joining ceramics to metals is difficult task due to differences in physical and thermal properties. [6]. A number of studies have been done in finding suitable ways in joining ceramics to metals. One of the common conventional methods is a brazing process. Conventional brazing is a joining process conducted within a closed furnace chamber with inert gas or vacuum at high temperature with the aid of brazing filler alloy [7, 8]. The drawbacks from this conventional technique are time consuming in which the cycle time of each product can exceed up to more than a day allowing time for slow cooling, close long hours monitoring, preliminary trial testing consumes longer time to make conclusion and whole workpiece is heated in the furnace while brazing. Besides that, the size and configuration limitation also urge the researchers to find alternative ways to overcome this issue. For certain workpiece, flux needs to be applied prior to brazing and removal of flux needed in post-brazing which adds more non-value added activities for the industry [9, 10].

Laser brazing is an advanced manufacturing process that has the potential to replace conventional brazing. Laser brazing uses a focus light beam to locally heat particular area that holds workpiece and preplaced brazing filler together. The thermal energy from the laser beams performed localized heating at brazed intended area [11]. The advantages of this advanced technique are precise dimension controlling in determining which part needs to brazed, rapid heating and minimum heat distribution on the whole workpiece. Laser brazing can be conducted in inert atmosphere which can be one of the influencing parameters to study for further improvement on the workpiece [12]. Laser brazing is flexible where wider range of parts and shapes can be carried out allowing for customization of products. Vacuum atmosphere does not necessarily offering higher productivity and flexibility [13]. Laser brazing between ceramic and metals face some difficulties due to the large mismatch on the thermal expansion coefficient and elastic modulus. This leads to poor wetting behaviour of liquid metals on ceramic to create adhesion joining. Therefore, the mechanical strength between the joint is low [3, 14]. To overcome this problem, active filler alloys consisting of elements such as titanium can be added during brazing to improve the wetting manners [3].

Artificial sapphire consisting of Al_2O_3 is transparent and single crystal ceramic with extreme hardness. The ability to withstand scratches and wear attracted many industrial applications such as in bioengineering, optics, electronics and aerospace [4, 15]. Inconel 600 is a nickel-based super alloy that is extensively used as parts in the aerospace industry because of their superior mechanical properties and excellent oxidation resistance at high temperatures. It makes the alloy suitable to be manufactured as components in high-temperature regions of aero engines, gas turbines and race car exhaust systems [16, 17]. On the other hand, Cusil ABA is a silver-copper base filler alloy with active element, titanium that helps to overcome the mismatch on the thermal expansion coefficient of ceramic and metals. Similar research were done to join sapphire to Inconel 600 through diffusion brazing [18] and conventional brazing with porous as a part of high temperature pressure sensor [4]. However, these techniques consumed longer time for joining the materials. There are some studies on the laser brazing of ceramic to metal, however significant findings on the brazability of sapphire to Inconel 600 using low power fiber laser are yet to be found. From this study, the feasibility of joining these materials were experimented and recommendations were made for future studies on laser brazing. Thus, the main objective of this study is to investigate the brazability of these ceramic-metal; sapphire and Inconel 600 through direct laser brazing using low power Yb-fiber laser by varying the laser welding speed and laser power.

2 Experimental Procedures

2.1 Materials and Sample Preparation

In the present work, sheets of Inconel 600 (20 mm × 20 mm × 1 mm) and artificial sapphire (diameter 15 mm × 1 mm) were chosen for lap laser brazing. Artificial sapphire has the chemical formula Al_2O_3, 85% transmission at 1070 nm wavelength. The melting temperature is 2030 °C for sapphire, meanwhile for Inconel 600, the solidus temperature is 1354 °C and liquidus temperature is 1413 °C. An active filler alloy, Cusil ABA with dimension 20 mm × 20 mm × 0.1 mm were placed in between Inconel 600 and sapphire to create bonding between them. Prior to laser welding, the samples were cleaned with acetone to remove any presence of surface contaminants. The chemical composition for Inconel 600, sapphire and filler alloy were given in Table 1.

2.2 Laser Brazing

A continuous wave Ytterbium (Yb)-fiber laser of 1070 nm wavelength with 300 W maximum power output (StarFiber 300, ROFIN-SINAR Technologies Inc., Germany) was used as heat source for the laser brazing. Laser beam hit the workpiece at perpendicular angle and circular (clockwise) movement as shown in Fig. 1 for 12 mm diameter. Inconel 600 was placed at the top followed by filler alloy and sapphire at the bottom. After few preliminary testing, the effects of laser power and welding speed for constant focal distance offset −7 (290 mm distance between scanner and workpiece with optimum distance is given as 346 mm by machine specification) were studied. Table 2 shows the detailed welding parameter for this study.

Table 1 Chemical composition of Inconel 600, sapphire and filler alloy

Materials	Element weight (%)									
	Ni	Cr	Fe	C	Si	S	Cu	Al_2O_3	Ti	Ag
Inconel 600	72 min	14.0–17.0	6.0–10.0	0.05–0.1	0.50 max	0.015 max	0.5 max	–	–	–
Sapphire	–	–	–	–	0.14	–	–	Balance	0.07	–
Cusil ABA	–	–	–	–	–	–	32.25	–	1.75	63

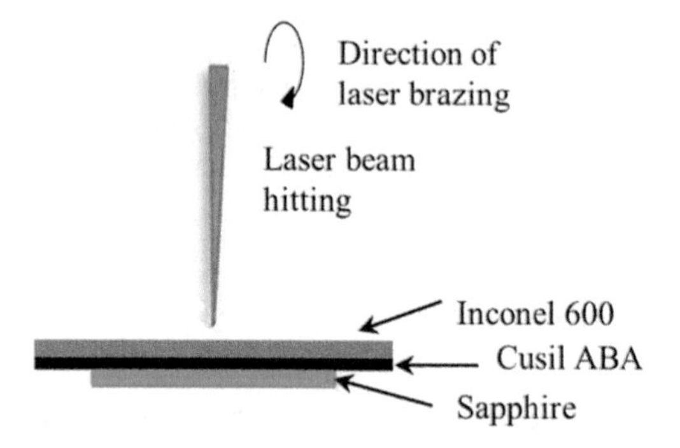

Fig. 1 Schematic diagram of laser brazing to create adhesion between samples

Table 2 Welding parameters used in the study

Parameter	Settings
Laser power	180, 185, 190 W
Welding speed	18, 19, 20 mm/s

2.3 *Microstructural Characterization*

Post brazing, samples were cold mounted with epoxy resin for 12 h and cut perpendicular towards the brazing direction for metallographic analysis as shown in Fig. 2. Initially, the workpiece were ground with sand paper grit 600, 800, 1200, 1500 and 2000. After fine grinding with sand paper grit 2500, the microstructures and the elemental composition of the weld bead cross-sections were observed using Optical Microscope (BX61, Olympus, Tokyo, Japan) and Scanning electron microscope (SEM, Phenom Pro X, Crest System (M) Sdn. Bhd., Eindhoven, The Netherlands) equipped with an energy dispersive X-ray (EDX, Phenom Pro X, Crest System (M)

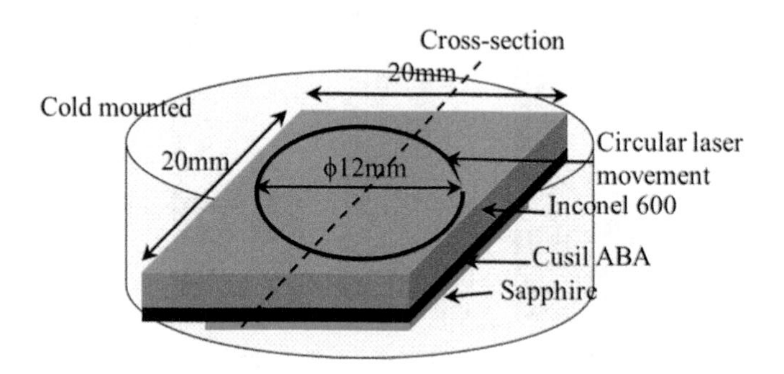

Fig. 2 Schematic diagram of cold mount samples for cross-section

Sdn. Bhd., Eindhoven, The Netherlands), respectively. The formation of intermetallic phases of the weld was characterized by X-ray powder diffractometer (XRD, PaNalytical Empyrean, DKHS Holdings (Malaysia) Bhd., Almelo, The Netherlands) using CuKα radiation in the 2θ diffraction angle range from 30° to 80°.

3 Results and Discussion

3.1 Metallographic and Microstructural Characterizations

During the laser brazing process, the Inconel 600 surface temperature rose and heated due to laser irradiation on the surfaces and subsequently the heat was transferred onto the brazing interface via conduction process [19]. Inconel 600 is on top of the configuration allowing the laser beam to hit the metal from top and the heat penetrated through the silver base filler alloy melting to make adhesive bonding with sapphire. Focal distance was adjusted at defocused position to allow shallow and wide conduction welding creation. Welding at defocused positions (focal length) produce different shapes of molten pools. The pools were shallow and wide called conduction welding comparing to deep penetration that gives keyhole welding [20]. Keyhole welding happens when a high energy density laser beam vaporize the surface of workpiece to form a hole that allows laser beam to penetrate into deep molten pool, meanwhile conduction welding happens at low power densities [21]. From the study, laser power 185 W at welding speed 18 mm/s and focal distance offset −7 gave an adhesion to the sapphire without breaking as shown in Fig. 3. Other parameters were not considered as successive joining because they were either broken or not joined at all.

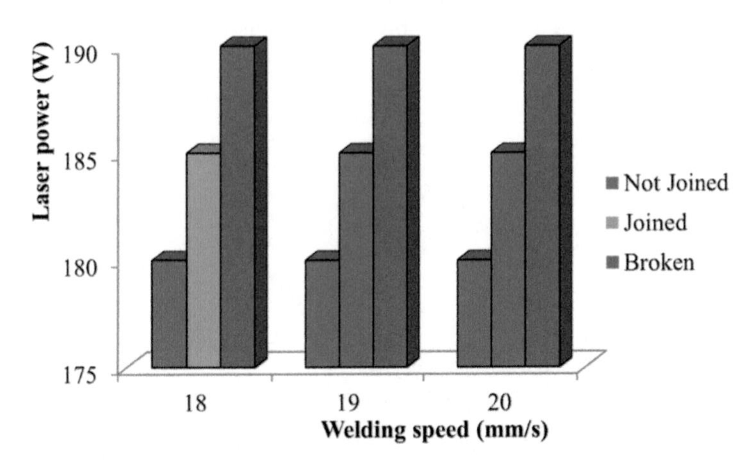

Fig. 3 Results obtained based on the appearance for adhesion

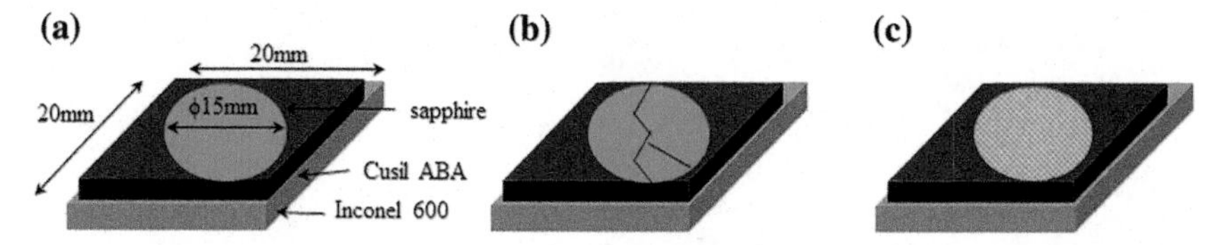

Fig. 4 Schematic diagram of samples **a** not joined **b** broken **c** joined

Figure 4 shows the schematic diagram of the samples that were not joined, broken and joined. Influences of power densities and heat input towards the workpiece determines the brazability of the samples. Higher power densities tends to scatter pieces meanwhile lower power densities does not help in melting the filler alloy and wet the samples to join them.

$$\text{Power density, } P_d = \frac{\mathrm{P}}{\pi \mathrm{r}^2} \tag{1}$$

$$\text{Laser beam radius, } \mathrm{r} = W_O \sqrt{1 + \frac{z}{Z_R}} \tag{2}$$

$$\text{Rayleigh range, } \mathrm{Z}_R = \frac{\pi W_O^2}{\lambda} \tag{3}$$

where W_O laser beam waist = 0.05 mm, $\lambda = 1.07$ nm (machine specification), Z_R calculated as 7.34 m, and z is differences between focal distance from optimum value (offset).

From Eqs. (1) to (3), the power density is calculated and heat input is calculated from Eq. (4). The power density for the joined sample is 398 W/mm^2 and heat input is 22.1 J/mm. For values of power density lower than this was not able to join the sample and higher power densities breaks the samples.

$$\text{Heat Input, H.I} = \frac{P_d}{\mathrm{v}}, \text{ v given as welding speed} \tag{4}$$

Figure 5 shows the SEM at the cross-sectioned sample to analyze the intermetallic layers that are formed [22]. The influence of silver to release titanium that formed an "ocean" structure had formed at the adhesive. However, the "ocean" structure was formed all over the adhesive bonding unlike conventional brazing where the structure was formed near to the interface of sapphire and Inconel 600 [4]. This has proved that laser brazing gives an even bonding between sapphire and Inconel 600. The EDX analysis proved the "ocean" structures were evenly formed as no distinguished layers or elements were found at the interface layer as shown in Fig. 6. Cu, Ti (from filler alloy), Cr (from Inconel 600) and Al (from sapphire) were distributed uniformly at the interface layer. Comparing to brazing microstructures, non-uniform reaction layer, mainly composed of Ti was formed at the interface brazed layers [23].

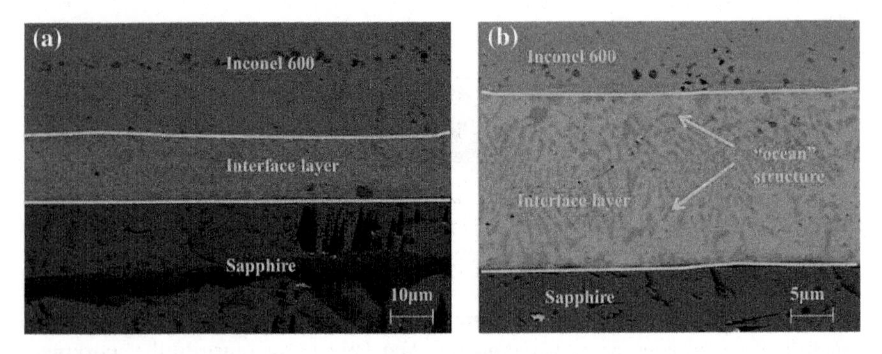

Fig. 5 SEM at cross-section of laser brazing between Inconel 600 and sapphire at **a** magnification of 1000× **b** magnification of 2000×

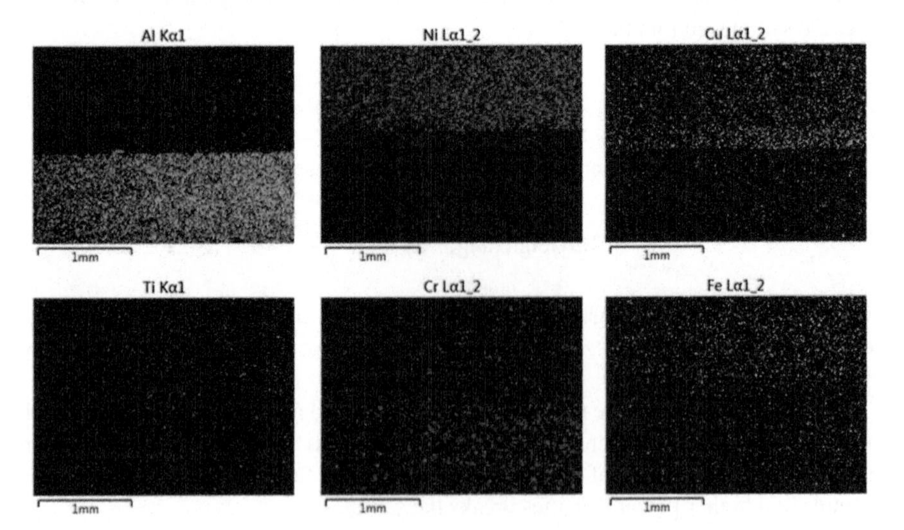

Fig. 6 EDX mapping along cross-section of laser brazing between Inconel 600 and sapphire

3.2 XRD Elemental Composition Analysis

Further investigations were conducted by XRD analysis on the detached sapphire and Inconel after laser brazing. Figure 7 shows the element present on both materials due to adhesion and intermetallic layers. Copper was not present at both the sides of materials because of the reaction of Ti from the active filler alloy [24]. After laser brazing, five phases were identified due to the reaction from the Cusil ABA towards Inconel 600 and sapphire. The identified phases are aluminum nickel (Al_3Ni_2 and $AlNi_3$), aluminum chromium ($AlCr_2$), chromium titanium (Cr_2Ti) and iron titanium (Fe_2Ti). Valette at. al and Zaharinie et al. studies also proved the presence of elements exist

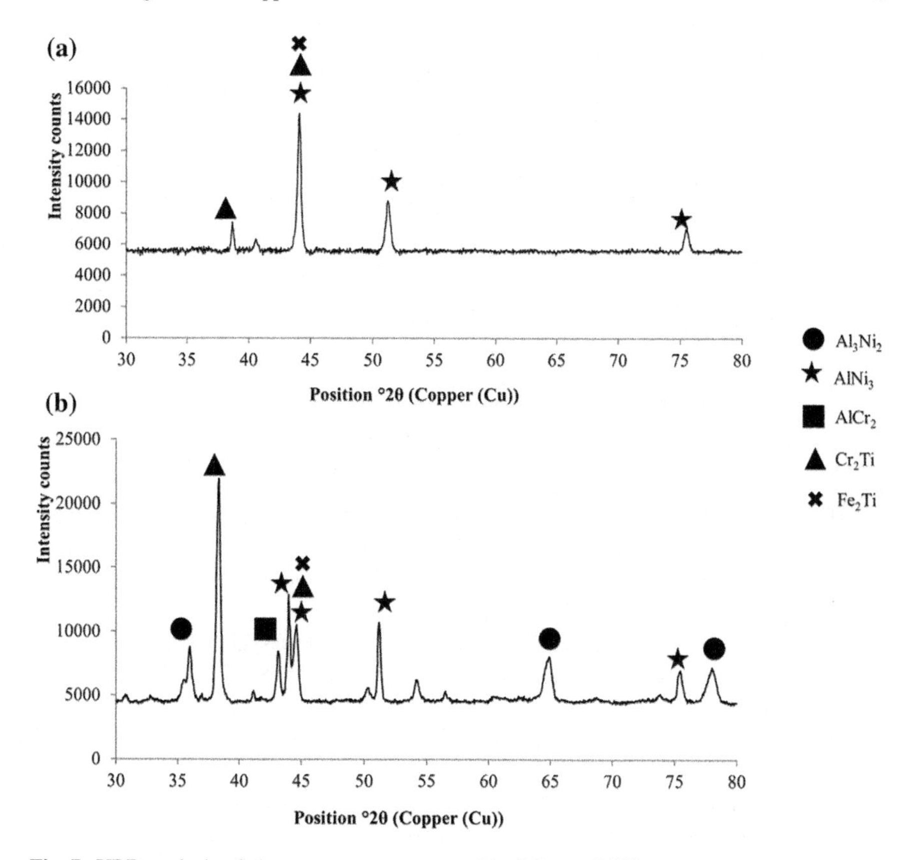

Fig. 7 XRD analysis of element present on **a** sapphire **b** Inconel 600

on the surface of bonding are Ti, Ni, Al, Cr and Fe. The formed chemical compounds are from the Ti reaction on the sapphire and Inconel surface. The potential of wetting reaction happening during laser brazing can be written as shown in Eqs. (5)–(9);

$$3Al + 2Ni \sim Al_3Ni_2 \tag{5}$$

$$Al + 3Ni \sim AlNi_3 \tag{6}$$

$$Al + 2Cr \sim AlCr_2 \tag{7}$$

$$Ti + 2Cr \sim Cr_2Ti \tag{8}$$

$$Ti + 2Fe \sim Fe_2Ti \tag{9}$$

From the XRD graphs, aluminum nickel ($AlNi_3$), iron titanium (Fe_2Ti) and chromium titanium (Cr_2Ti) are the only phases found on sapphire. Meanwhile, additional phases such as aluminum chromium ($AlCr_2$) and aluminum nickel (Al_3Ni_2) that were found on Inconel 600 surface proving intermetallic layers are easily formed on metals compared to ceramics. During conventional brazing, Ni element from Inconel 600 dissolves and reacts well with sapphire with the aid of brazing alloy forming Ti and Ni compounds. Some phases of Ti–Ni compounds can exhibit detrimental effect on the bonding and this can be avoided during laser brazing [23, 24]. The EDX results indicated Ti–Ni compound were not formed during laser brazing.

4 Conclusions

Through this study laser brazing between metal has been proven. Inconel 600 and ceramic, artificial sapphire had successfully created an adhesion to bond the materials. The observation from the joint is revealed as below:

(a) Feasibility study has proven that adjusting the laser power, welding speed and defocusing focal distance can join sapphire to Inconel 600 within very short period overcoming the conventional brazing technique that requires longer duration.
(b) The optimum parameters, laser power 185 W at welding speed 18 mm/s and focal distance offset −7 gave a sufficient adhesion to the sapphire without breaking it.
(c) Intermetallic layers were easily formed on metal surfaces compared to ceramic surfaces.
(d) "Ocean" structures were formed evenly at the interface layer proving laser brazing created uniform bonding comparatively to conventional brazing.

Acknowledgements The authors greatly acknowledge the University of Malaya, Kuala Lumpur for providing the necessary facilities and resources for this research. The authors acknowledge the contributions of SEGi University, Kota Damansara, Petaling Jaya for the fund given for conference and publication for this study.

References

1. Jacobson DM, Humpston G (2005) Principles of Brazing. ASM International
2. Sechi Y, Tsumura T, Nakata K (2010) Dissimilar laser brazing of boron nitride and tungsten carbide. Mater Des 31(4):2071–2077
3. Rohde M, Südmeyer I, Urbanek A, Torge M (2009) Joining of alumina and steel by a laser supported brazing process. Ceram Int 35(1):333–337
4. Zaharinie T, Moshwan R, Yusof F, Hamdi M, Ariga T (2014) Vacuum brazing of sapphire with Inconel 600 using Cu/Ni porous composite interlayer for gas pressure sensor application. Mater Design 54:375–381
5. Indacochea JE, McDeavitt S, Billings GW (2001) Interface interactions of dissimilar materials at elevated temperatures. Adv Eng Mater 3(11):895–901
6. Do Nascimento R, Martinelli A, Buschinelli A (2003) Review article: recent advances in metal-ceramic brazing. Cerâmica 49(312):178–198

7. Lippmann W, Knorr J, Wolf R, Rasper R, Exner H, Reinecke A-M, Nieher M, Schreiber R (2004) Laser joining of silicon carbide—a new technology for ultra-high temperature resistant joints. Nucl Eng Des 231(2):151–161
8. Haferkamp E, Bach F-W, Von Alvensleben F, Tuan AM, Kreutzburg K (1996) Laserstrahlloten von Metall-Keramik-Verbindungen. Schweissen + Schneiden 11
9. Schwartz M (1994) Brazing: for the engineering technologist. Springer, Netherlands
10. Roberts P (2013) Industrial Brazing practice, 2nd edn. CRC Press
11. Schwartz MM (2003) Brazing, 2nd edn. ASM Int
12. Messler RW (2004) Joining of Materials and Structures: From Pragmatic Process to Enabling Technology. Elsevier
13. Ion J (2005) Laser processing of engineering materials: principles, procedure and industrial application. Elsevier Science
14. Zaharinie T, Yusof F, Hamdi M, Ariga T, Fadzil M (2013) Microstructure Analysis of Brazed Sapphire to Inconel® 600 Using Porous Interlayer. Weld J
15. Wang G, Zuo H, Zhang H, Wu Q, Zhang M, He X, Hu Z, Zhu L (2010) Preparation, quality characterization, service performance evaluation and its modification of sapphire crystal for optical window and dome application. Mater Des 31(2):706–711
16. Chen HC, Pinkerton AJ, Li L (2011) Fibre laser welding of dissimilar alloys of Ti-6Al-4V and Inconel 718 for aerospace applications. Int J Adv Manuf Tech 52 (9–12):977-987. https://doi.org/10.1007/s00170-010-2791-3
17. Janasekaran S, Tan A, Yusof F, Abdul Shukor M (2016) Influence of the overlapping factor and welding speed on t-joint welding of Ti6Al4V and Inconel 600 using low-power fiber laser. Metals 6(6):134
18. Ishihara T, Sekine M, Ishikura Y, Kimura S, Harada H, Nagata M, Masuda T (2005) Sapphire-based capacitance diaphragm gauge for high temperature applications. In: The 13th international conference on IEEE, solid-state sensors, actuators and microsystems. Digest of technical papers. TRANSDUCERS'05, pp 503–506
19. Nath AK, Sridhar R, Ganesh P, Kaul R (2002) Laser power coupling efficiency in conduction and keyhole welding of austenitic stainless steel. Sadhana-Acad P Eng S 27:383–392. https://doi.org/10.1007/bf02703659
20. Khalid M, Hafez KS (2009) Fiber laser welding of AISI 304 stainless steel plates. Q J Jpn Weld Soc 41(36):69s–73s
21. Walsh C (2002) Laser welding–literature review. Materials Science and Metallurgy Department, University of Cambridge, England Retrieved from http://www.msm.cam.ac.uk/phasetrans/2011/laser_Walsh_review.pdf. Accessed 22 Aug 2016
22. Mandal S, Ray AK, Ray AK (2004) Correlation between the mechanical properties and the microstructural behaviour of Al_2O_3–(Ag–Cu–Ti) brazed joints. Mater Sci Eng, A 383(2):235–244
23. Zaharinie T, Yusof F, Fadzil M, Hamdi M, Ariga T (2014) Microstructural analysis of brazing sapphire and Inconel 600 for sensor applications. Mater Res Innovations 18(sup6):S6-68–S66-72
24. Valette C, Devismes M-F, Voytovych R, Eustathopoulos N (2005) Interfacial reactions in alumina/CuAgTi braze/CuNi system. Scripta Mater 52(1):1–6

A Review on Underwater Friction Stir Welding (UFSW)

Dhanis Paramaguru, Srinivasa Rao Pedapati and M. Awang

Abstract Underwater Friction Stir Welding (UFSW) is noted as an advanced technique in welding field which is a really new and emerging technology in recent years. In the present paper, a brief explanation on introduction to the Underwater Friction Stir Welding (UFSW) technique along with a review on the latest researches have been made. The review is designed based on joint strength analysis, thermal distribution analysis, microstructural analysis, process modelling and computing techniques, effect of thermal boundary condition in UFSW, effect of process parameters, defects in UFSW and dissimilar welds. The applications of UFSW have also been discussed, along with a detailed description of advantages and limitations of UFSW technique. Lastly, the possible future research exploration has been proposed.

1 Introduction

Friction Stir Welding (FSW) is a comparatively modern, besides unique form of solid-state joining method that applies a non-consumable tool to weld two facing workpieces without melting the workpiece material. This technique is invented and experimentally proven by Wayne Thomas at The Welding Institute (TWI) of United Kingdom (UK) in December 1991. Also, TWI held patents in this operation method which is the first well detailed procedure. At first, the FSW method was observed as a "laboratory" inquisitiveness, however then it was established as a technique which provide virtuous advantages in products fabrication [1, 2]. This technique is rapidly employed as an attractive operation to fabricate lightweight products in area like aerospace, aircraft, marine, automobile, railway and food processing industries for about a decade [3, 4]. It is able to join various types of materials such as metals, ceramics, polymers and etcetera [5, 6]. In the past few years, UFSW has been introduced as a new solid-state welding technique. It takes place at temperatures lower

D. Paramaguru · S. R. Pedapati (✉) · M. Awang
Universiti Teknologi PETRONAS, Seri Iskandar, Perak, Malaysia
e-mail: Srinivasa.pedapati@utp.edu.my

M. Awang (ed.), *The Advances in Joining Technology*, Lecture Notes in Mechanical Engineering, https://doi.org/10.1007/978-981-10-9041-7_6

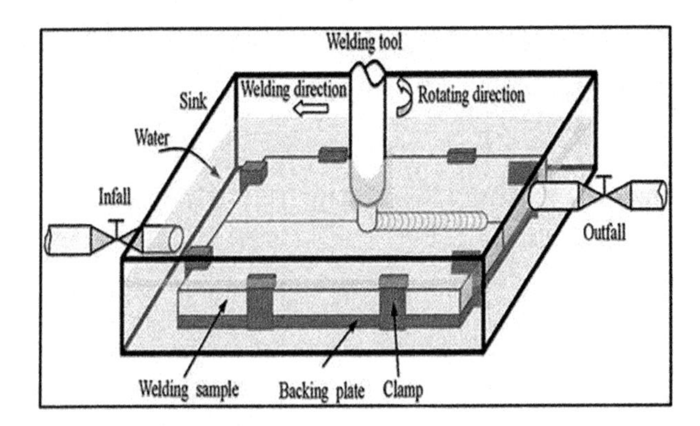

Fig. 1 Schematic diagram of Underwater FSW [5]

than the melting point of the material, where rotating tool shoulder rubs between the alloy surface of workpieces under the water, as illustrated in Fig. 1. It generates enough heat due to friction to melt the workpiece beneath that benefits the tool pin to stir the melted material and cause plastic deformation to produce a weld joint.

Besides believed as one of the innovative welding practices in the present era, UFSW also helps to avoid defects such as shrinkage, cracking, solidification, splatter, cracking, embrittlement and porosity to occur [7]. By the same token, this technique does not involve any shield protective gas or electrodes to produce an arc like some different joining methods during the welding process. It makes this process cheaper and requires less energy. UFSW also produce fine defined differences in grain size between different regions, great quality in weld joint in a short cycle time and improve mechanical properties.

2 Studies on Underwater Friction Stir Welding (UFSW)

2.1 Joint Strength Analysis

Thomas, [8] expounded in detail a research work on underwater welding technique to increase the ultimate tensile strength of friction stir weld using AA6061. Fratini et al. [9] observed a successful in-process heat treatment by water streaming above the AA7075-T6 aluminium alloy plates during FSW. The goal of this paper is to study the possibility to improve the joint performances. Comparably, another two different welded joints were carried out in normal FSW conditions: free air and forced air. Both metallurgical and mechanical examinations were developed on the welded joints. It was proven that substantial enhancement of ultimate tensile strength was obtained in all the high, medium and low level of heat input controlled by welding parameters variation. The water cooling effect significantly reduced the soften zones induced to develop the weld mechanical properties.

Nelson et al. [10] clearly defined a literature review on effective influence of cooling rate from thermal exposure on the weld performance of AA7075-T7351 through FSW technique. The outcome of this paper indicated that 7075-T7351 is a quench sensitive alloy because of more quick natural aging response and enhanced mechanical properties. The natural aging response was evaluated through transverse tensile properties and micro-hardness study. Most importantly, the water cooling conditions approximately increased the tensile properties by 10% above normal FSW.

Zhao et al. [11] also states that tensile strength of Underwater FSW method reached 75% of base metal and the elongation is comparably greater than the normal FSW joint. In this research, ultra-high strength spray formed AA7055 is welded using FSW method in air (normal FSW) and underwater, respectively. This Underwater FSW process was applied to reduce the heat input and increase the joint properties by varying welding temperature history. A better performance of underwater welded joint was illuminated through reduced residual stress and minimum thermal cycle curve. Furthermore, the hardness, tensile strength and plasticity of Underwater FSW joint were improved compared to normal FSW joint properties. The underwater joint also produced microstructure with fine grained characteristic which diminished "S line" type of defect and has clear borderline in the middle of WNZ and TMAZ, while reduced HAZ.

Liu et al. [12] studied Underwater Friction Stir Welding (UFSW) for AA2219-T6 to clarify the enhanced value in tensile strength compared to normal FSW (in air) joint. These joints were cut into three different layers (lower layer, middle layer and upper layer) to investigate the homogeneity of mechanical properties of the joint. Tensile strength of all these three layers of the joint was improved as a result of this UFSW technique. Compared to upper layer, the middle and lower layers were recorded great amount of strength improvement which leading to an increase in joints mechanical properties, along with improvement in minimum hardness value of the joint. This study also verified that the effect of water cooling is the fundamental cause for the UFSW joint to increase the strength. However, this study did not include weld temperature distribution.

2.2 *Thermal Distribution Analysis*

Liu et al. [13] again focused their research on UFSW of 2219 aluminium alloy to further advance the mechanical properties of the joint by varying welding temperature history. This research is able to discover that external water cooling action in UFSW developed the normal FSW joint tensile value from 324 to 341 MPa. Nevertheless, plasticity of the weld is weakened. It also concluded that UFSW joint have a tendency to fracture in the middle of Thermal Mechanically Affected Zone (TMAZ) and Weld Nugget Zone (WNZ), on the Advancing Side (AS) during tensile test.

Sakurada et al. [14] In their investigation, were the first who used submersion, focused on the high-speed rotating cylindrical sample with different welding conditions in Underwater FSW of 6061 aluminium alloy. The experimental results proved that underwater joint generate less peak temperature compared to normal FSW joint. The softening ratio and the width of softening area were reduced through Underwater FSW process while enhanced the joint efficiency (obtained 86%), in withal.

Hofmann and Vecchio [15] investigated and revealed that additional grain refinement was achieved through Submerged friction stir processing (SFSP) condition because of faster cooling rate. The grain sizes were predicted using boundary migration model along with recorded thermal distribution in the stirred material. Besides, the microstructures were characterized using Transmission Electron Microscopy (TEM).

2.3 *Microstructural Analysis*

Zhang et al. [16] conducted experimental study to discover the basic justification for the mechanical properties enhancement in the Heat-Affected Zone (HAZ) by UFSW of AA2219-T6. The experimental observations through microstructural analysis exposed that the hardness of the HAZ can be enhanced by UFSW method due to the narrowing of precipitate free zone and the shortening of hardened precipitate level. The variations in welding thermal distribution by the effect of water cooling treatment is the basic cause for differences of mechanical properties and microstructures in the HAZ of UFSW joint.

Hosseini et al. [17] deliberated the effect of Underwater FSW approach on the microstructure and mechanical characterization of the joint, in comparison with normal FSW (in air). This research has an objective to diminish the weakening of joints mechanical properties using Ultra Fine-Grained strain hardenable 1050 Aluminium Alloy. With respect to the normal FSW technique, the evaluated microstructure of the Underwater FSW condition using Transmission electron microscopy and X-ray diffraction examinations exposed smaller final grains and sub-grain sizes in stir zone through slowing in grain growth rate and less softening occurred in stir zone (SZ). Moreover, the Underwater FSW is also resulted tensile and yield strength increment, improved super-plasticity tendency of the material, besides narrowed the HAZ.

Wang et al. [18] studied the metallurgical and mechanical characterization through Underwater FSW joints of spray forming 7055 aluminium alloy (T6). The microstructure of joints was examined through Optical Microscope (OM) and Scanning Electron Microscope (SEM), while Energy-dispersive X-ray spectroscopy (EDS), X-ray diffraction (XRD), Differential Scanning Calorimetry (DSC) and Transmission Electron Microscopy (TEM) were used to analyse the strengthening phase's developments. It is indicated that the hard-etched area was completely removed (with 'W' shaped distribution of hardness) from microstructure of UFSW joint in cooperation with normal FSW, as shown in Fig. 2. The water cooling treatment of UFSW process enhanced the thermal cycle of welding while influence the strengthening mechanism

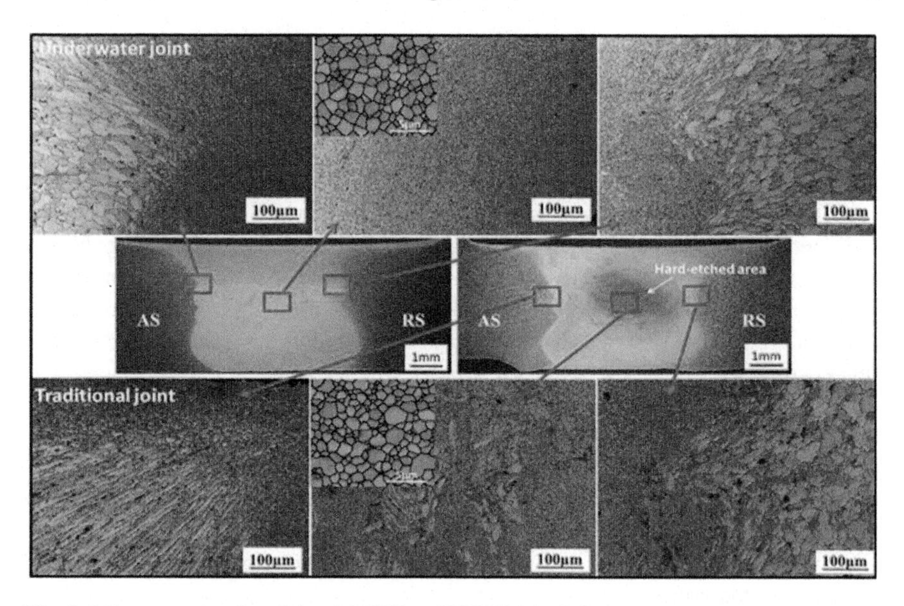

Fig. 2 Microstructure of traditional (FSW) and UFSW joint [18]

and microstructure of the joints. Defect free joint was reported with twice the elongation (1.96%) and enhanced the tensile strength as 406.06 MPa (~30%), compared to normal FSW joint.

Hofmann and Vecchio [19] conducted their research on submerged friction stir processing (FSP-modification of submerged FSW) using Al-6061–T6 material as an alternative and advance practice to produce ultrafine-grained bulk materials in stir zone (SZ) by great plastic deformation, compared to normal FSP (in air). It is attributed to the less extent of thermal distribution to the weld material during the welding operation. Thermocouples and Transmission Electron Microscopy (TEM) results were used to analyse the outcome of this paper.

2.4 *Process Modelling and Computing Techniques*

Fratini et al. [20] further studied an integrated numerical and experimental exploration by effects of an in-process water cooling action using FSW technique for AA7075-T6 butt joints. The temperature distribution, tensile strength and microstructure of the welded joint were observed, along with an expanded finite element to interpret specimens obtained under various process conditions. Despite the analysed temperature histories, in-process water cooling treatment was increased joint strength, diminished destructive effect on nugget zone, besides reduced the material softening and thermal flow adjacent to the tool.

Zhang et al. [21] represents the 3D thermal modelling as an advanced exploration on temperature histories of Underwater FSW technique by using the mathematical modelling approach based on heat transfer model. Experimental results are also analysed to validate efficiency of the thermal model, while disclosed good agreement with the calculated results. Mathematical model was examined the vaporizing aspect of water to clarify the conditions of boundary, while considering the material's temperature dependent characteristics. It was revealed that welding thermal cycles in different zones and area of high-temperature distributions are significantly reduced via Underwater FSW technique. Compared to normal FSW, the utmost peak temperature of UFSW joint was minimized, even though the shoulder surface heat flux is greater.

Zhang and Liu [22] further examined UFSW and developed a mathematical model using 2219-T6 aluminium alloy to optimize the welding parameters for maximum tensile strength. Highest tensile value of 360 MPa was obtained through UFSW operation and it was comparably 6% greater than the highest tensile value of FSW operation in air. This study concluded that the basic reasons for increment in tensile strength through UFSW were microstructural developments and controlling of temperature histories.

Sree Sabari et al. [23] examined the microstructural appearance and mechanical characteristics of UFSW joint with maximum strength, armour grade 2519-T87 aluminium alloy. For comparison, similar material joints were made by normal FSW (in air). The study composed of tensile test, microstructure examination, micro-hardness, fracture surface analysis and thermal analysis of all joints. A finite element analysis is used to evaluate the width of Thermo-Mechanically Affected Zone (TMAZ) and temperature distribution. The outcomes were compared with results from experimental analysis. It is concluded that Underwater FSW experienced higher peak temperature (547 °C), higher cooling rate and higher temperature gradient compared to normal FSW joint attributed to heat absorption ability of the water cooling system. Additionally, UFSW also reduced the width of TMAZ as the weaker zone and over aging of HAZ which substantially increase the tensile properties of the joint where the joint efficiency was improved by 60%.

2.5 *Effect of Thermal Boundary Condition in UFSW*

Fu et al. [24] conducted their study and then investigated the micro-hardness distribution, weld thermal cycles and tensile strength for UFSW of 7050 aluminium alloy. In this study, produced weld joints under hot and cold water, as well as in air for comparison. The outcomes indicated that maximum temperature was recorded during welding process in normal FSW (in air), after welding process in hot and cold water. It has been discovered that joint's retreated side accounted maximum temperature in contrast to advanced side and it has been recommended that weld joint under hot water is the best compared to other two conditions where it improved the mechanical characteristics of the weld. This results in a ratio of 150% elongation and 92% ultimate tensile strength. The fracture positions were situated in HAZ area (lowest micro-hardness location). Width of minimum hardness zone is changing accordingly with the ambient conditions.

Darras and Kishta [25] In their research, friction stir processing technique using three different conditions: normal FSW (in-air), under-hot-water and underwater-at-room-temperature were effectively compared. This experiment was conducted using AZ31B-O Magnesium alloy. It is analysed and supported that Underwater FSW generated better grain refinement, reduced both time spent on some reference temperature and peak temperature, besides minimized porosity and increased the formability of alloy. Particularly, the formability of this alloy is enhanced by UFSW in hot water.

Upadhyay and Reynolds [26] In their work, theoretically clarified the influence of varying the thermal boundary condition and control variables on friction stir welding process using AA7050-T7 material which were examined in-air (normal FSW), underwater, and under sub-ambient temperature (−25 °C) conditions. From the study, it was possible to conclude that welding underwater compared to normal FSW expressively lessened the size of nugget grain, enhanced amount of cooling in the HAZ, increased the hardness of weld nugget, decreased probe temperature, besides increasing power consumption and torque. The ultimate tensile strength of Underwater FSW in all range of parameters were presented good improvements along with elongation of the joint. Yet, it is also justified that sub-ambient temperature (−25 °C) condition does not contribute a consequential benefit in contrast to the underwater welding at ambient temperature.

2.6 *Effect of Process Parameters*

Liu et al. [27] clarified the influence of speed of welding from 50 to 200 mm/min on the efficiency of underwater friction stir welded joints. This investigation used 2219 aluminium alloy with fixed rotational speed equivalence to 800 rpm. It resulted in the weakening on the precipitate degradation in TMAZ and HAZ with increasing speed of welding. Subsequently, it leads to increase in lowest hardness value and reduced the softening region. The study also stated the joints fracture features are basically reliant on the speed of welding, and increasing the speed of welding increases tensile strength of the defect-free joints. Nevertheless, the temperature range applied in the investigation was restricted less than room temperature.

Zhang et al. [28] additionally explored the impact of rotation speed on mechanical properties of UFSW weld using AA2219-T6 with constant speed of welding. The tensile properties, microstructural characteristics and hardness distributions of the joints were illuminated through this investigation. The joint tensile properties were extremely sensitive to the rotational speed where it was dramatically increased from rotational speed of 600–800 rpm, and later reached a plateau in a large range of rotation speed. Thenceforth, notable reduction in tensile properties was attained due to void defects formations in the SZ. Escalate in dislocation density as well as in grain size of SZ were reported with increasing of rotational speed, which slowly increased the SZ hardness. At higher speed of rotation, the defect-free joints fracture locations changed to the HAZ or Thermal-Mechanically Affected Zone (TMAZ) as the hardness increased in the SZ. Meanwhile, the welded joint was fractured in the SZ at lower speed of rotation.

Kishta and Darras [29] analysed and presented the impact of different parameters of process, namely rotational speed and translational speed, on Underwater FSW of 5083 marine-grade aluminium alloy. The outcomes of UFSW joints were compared with normal FSW (in air) joints, in withal. The void fractions, tensile properties, micro-hardness, thermal histories, and the process power consumption were comprehensively discussed. Investigation concluded that UFSW has produced good quality welds by higher rotational speed due to excellent thermal capacity of water, peak temperature decrement as well as cooling rate increment. The fraction of void-area in the SZ of Underwater FSW joint was decreased significantly nearly one-third of the base material. The maximum micro-hardness value was recorded in the SZ, while the UFSW joint elongation upsurge to almost two times the elongation of the base material.

Abbas et al. [30] examined the effect on weld quality of sample is investigated through the relation between the tool profile, welding speed and angle of tool inclination of Underwater FSW process enveloped for 6061 aluminium plate. After a brief description of operational principal of friction stir welding, the experiment setup is illustrated in detail together with clamping structure, welding tool properties, material properties and process parameters. The weld quality is evaluated through microstructure analysis, tensile strength test and Vickers hardness. Microstructure analysis disclosed that very few amounts of porosity is detected, while a good joining is obtained with no voids and cracks by UFSW technique. The mechanical properties of UFSW is increased approximately 20% compared to FSW. The Taguchi optimization technique has been used to analyse the optimized parameter with Mini Tab 16 software.

2.7 Defects in UFSW

Zhang and Liu [31] expounded in detail the outcomes of an experimental investigation, done on the UFSW of 2219-T6 aluminium alloy. Preceding research studies showed the influence of process parameters: speed of rotation and speed of welding on the quality of UFSW. Hence, the investigation objective is to comprehend the characteristics of welding defects and formation mechanisms of the welding defects in UFSW joints through examining the material flow patterns, as shown in Fig. 3. It is an attractive method to illustrate the results of experiment and it is really beneficial to provide guidance for process optimization. At low and high speed of rotation yield welding flaws during UFSW process. At high rotation speed and low speed of welding, defect formed is influenced by the high amount of extruding reflux of stir zone material on the AS. While at high rotation speed and high speed of welding, a

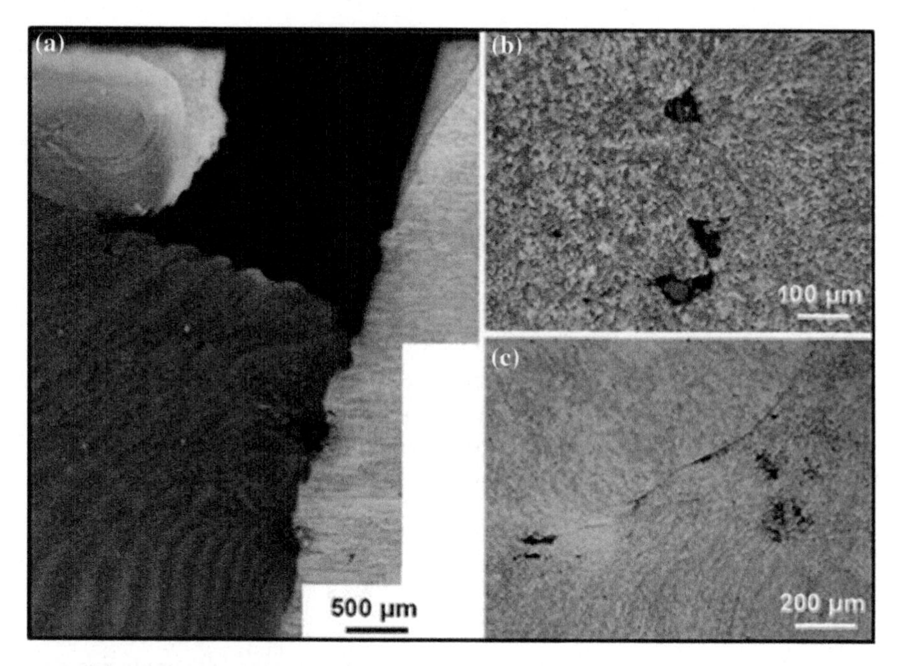

Fig. 3 Welding defects formed under different process parameters: **a** 800 rpm–200 mm/min, **b** 1000 rpm–300 mm/min, **c** 1400 rpm–100 mm/min [31]

great volume of material from TMAZ is pulled into the pin hole reduces the quantity of stir zone material that flows back to AS. Consequently, it formed groove and void type defects in the joints. Also, the low rotational speed flaws are usually found at the TMAZ and SZ boundary on the AS.

2.8 Dissimilar Weld Joints by UFSW

Mofid et al. [32] explored the effect of dissimilar welds in Underwater Friction Stir Welding (UFSW) using 5083 Aluminium alloy and AZ31C–O Magnesium alloy. The outcome revealed that submerged FSW technique enhanced the fine grained welds and lessen the development of intermetallic phases due to lower temperature attained. It impacts the joint mechanical characteristics substantially. Nevertheless, normal FSW (in air) produced great peak temperature in SZ with great amount of joint hardness in the centre, in contrast to UFSW joint.

Mofid et al. [33] discussed the impact of submerged welding using liquid nitrogen and underwater on the grain refinement in dissimilar materials of AA5083 H34 and AZ31 (Mg alloy). For comparison, three different environments: in air, water, and liquid nitrogen using parameters of 400 rpm and 50 mm/min were applied. Results of microstructure, Scanning Electron Microscopy (SEM), Energy-Dispersive

Spectroscopy (EDS), temperature profile, micro-hardness and tensile testing were systematically analysed. It is concluded that, Submerged FSW method suppresses formation of brittle interatomic compounds due to lower peak temperature.

3 Applications, Advantages and Limitations of UFSW

3.1 Applications of UFSW

The most significant applications of Underwater FSW method are building large sized ships beyond the capacity of facilities in existing harbours, maintenance and repairing works of ships, temporary reconstruction works due to unexpected accidents of ships, recover containers sunk in the sea and offshore construction for pipelines [34].

3.2 Advantages of UFSW

The good qualities and advantages of Underwater FSW method are producing great quality and decent joint in limited cycle time, not requires filler metals and does not produce shielding gasses, able to weld most of the common metals, not difficult to operate and good operation flexibility in all positions with simple automated function. Moreover, this UFSW method is also able to weld numerous types of dissimilar materials, generate fine-grained forged joint by eliminate weld inclusions or weld dilution, produce reliable welds while consume less energy during the joining process [5, 34].

3.3 Limitations of UFSW

The disadvantages of Underwater FSW method are the inspection process for welded joints by underwater friction stir welding (UFSW) technique may be harder compared to normal FSW, promising a better quality of UFSW joints much difficult and risk of not detecting the defects properly might occur. Besides, requires quite expensive machines as well as the machine tools [34].

4 Future Scope of UFSW

Preceding substantial researches have been constructed to advance the control strategies and process performance of underwater FSW method. Nevertheless, there are many conflicts to solve where the UFSW exploration should be focusing detailed research on properties of welded material and process optimisation. In addition, research on UFSW must develop the potential usage of robot manipulator for underwater FSW joints of complex geometry to improve the automation of UFSW joining and examination process, aside from developing the application of underwater FSW for large and complex scale of any structures. Furthermore, this technique should explore in detail on thermal management in terms of both closed-loop temperature control and thermal boundary condition modification, while improving the in-process weld quality assurance, increasing the application of UFSW to a wider range of engineering materials and improving the control techniques for continuous welding [34].

5 Conclusions

UFSW is really an innovative and developed technique of joining process in this present era. From the above literature review, it is disclosed that very few investigations are done based on underwater FSW method. It is acknowledged as an advanced welding technique and very little number of studies have applied the optimization technique in the field of UFSW. UFSW is not fully examined yet. By focusing the future research on it, can really enhance and attain good weld joint with economical, environment friendly and safe welding condition.

Acknowledgements The authors would like to acknowledge the financial support from the Ministry of Higher Education (MOHE) Malaysia through Fundamental Research Grant Scheme (FRGS) grant Ref. No. FRGS/1//2015/TK03/UTP/02/6.

References

1. Nandan R, DebRoy T, Bhadeshia H (2008) Recent advances in friction-stir welding-process, weldment structure and properties. Prog Mater Sci 53:980–1023
2. TWI Ltd., "TWI Group websites," The Welding Institute, 29 March 2017. [Online]. Available: http://www.twi-global.com/capabilities/joining-technologies/friction-processes/friction-stir-welding/
3. Ibrahim NT, Mustafa D, Hasan O, Ahmet Y (2010) Optimizations of friction stir welding of aluminum alloy by using genetically optimized neural network. Int J Adv Manuf Technol 48:95–101
4. Dolby R, Sanderson A, Threadgill P (2001) Recent developments and applications in electron beam and friction technologies. In: 7th international Aachen welding conference, Germany
5. Tulika G, Priyank M, Varun S, Chirag J, Prateek G (2014) Underwater friction stir welding: an overview. Int Rev Appl Eng Res 4(2):165–170
6. Thomas WM, Nicholas ED, Needham JC, Murch MG, Temple-Smith P, Dawes CJ (1991) Friction-stir butt welding. United Kingdom Patent PCT/GB92102203
7. Supraja Reddy B (2016) Review on different trends in friction stir welding. Int J Innov Eng Technol 7(1):170–178

8. Thomas B (2009) On the immersed friction stir welding of AA6061-T6: a metallurgic and mechanical comparison to friction stir welding. Nashville, Tennessee, May 2009, thesis for the degree of master of science in mechanical engineering
9. Fratini L, Buffa G, Shivpuri R (2009) In-process heat treatments to improve FS-welded butt joints. Int J Adv Manuf Technol 43:664–670
10. Nelson TW, Steel RJ, Arbegast WJ (2003) In situ thermal studies and post-weld mechanical properties of friction stir welds in age hardenable aluminium alloys. Sci Technol Weld Join 8(4):283
11. Zhao Y, Wang Q, Chen H, Yan K (2014) Microstructure and mechanical properties of spray formed 7055 aluminum alloy by underwater friction stir welding. Mater Des 56:725–730
12. Liu H, Zhang H, Yu L (2011) Homogeneity of mechanical properties of underwater friction stir welded 2219-T6 aluminum alloy. ASM Int 4
13. Liu HJ, Zhang HJ, Huang YX, Yu L (2010) Mechanical properties of underwater friction stir welded 2219 aluminum alloy. Trans Nonferrous Met Soc China 20:1387–1391
14. Sakurada D, Katoh K, Tokisue H (2002) Underwater friction welding of 6061 aluminum alloy. J Jpn Inst Light Metals 52(2):2–6
15. Hofmann DC, Vecchio KS (2007) Thermal history analysis of friction stir processed and submerged friction stir processed aluminum. Mater Sci Eng, A 465:165–175
16. Zhang H, Liu H, Yu L (2012) Effect of water cooling on the performances of friction stir welding heat-affected zone. ASM Int JMEPEG 21:1182–1187
17. Hosseini M, Danesh Manesh H (2010) Immersed friction stir welding of ultrafine grained accumulative roll-bonded Al alloy. Mater Des 31:4786–4791
18. Wang Q, Zhao Z, Zhao Y, Yan K, Liu C, Zhang H (2016) The strengthening mechanism of spray forming Al–Zn–Mg–Cu alloy by underwater friction stir welding. Mater Des 102:91–99
19. Hofmann DC, Vecchio KS (2005) Submerged friction stir processing (SFSP): an improved method for creating ultra-fine-grained bulk materials. Mater Sci Eng, A 402:234–241
20. Fratini L, Buffa G, Shivpuri R (2010) Mechanical and metallurgical effects of in process cooling during friction stir welding of AA7075-T6 butt joints. Acta Mater 58:2056–2067
21. Zhang HJ, Liu HJ, Yu L (2013) Thermal modeling of underwater friction stir welding of high strength aluminum alloy. Trans Nonferrous Met Soc China 23:1114–1122
22. Zhang H, Liu H (2013) Mathematical model and optimization for underwater friction stir welding of a heat-treatable aluminum alloy. Mater Des 206–211
23. Sree Sabari S, Malarvizhi S, Balasubramanian V, Madusudhan Reddy G (2016) Experimental and numerical investigation on under-water friction stir welding of armour grade AA2519-T87 aluminium alloy. Defence Technol 12:324–333
24. Fu RD, Sun ZQ, Sun RC, Li Y, Liu HJ, Liu L (2011) Improvement of weld temperature distribution and mechanical properties of 7050 aluminum alloy butt joints by submerged friction stir welding. Mater Des 4825–4831
25. Darras B, Kishta E (2013) Submerged friction stir processing of AZ31 magnesium alloy. Mater Des 47:133–137
26. Upadhyay P, Reynolds A (2010) Effects of thermal boundary conditions in friction stir welded AA7050-T7 sheets. Mater Sci Eng, A 527:1537–1543
27. Liu H, Zhang H, Yu L (2011) Effect of welding speed on microstructures and mechanical properties of underwater friction stir welded 2219 aluminum alloy. Mater Des 1548–1553
28. Zhang H, Liu H, Yu L (2011) Microstructure and mechanical properties as a function of rotation speed in underwater friction stir welded aluminum alloy joints. Mater Des 32:4402–4407
29. Kishta EE, Darras B (2014) Experimental investigation of underwater friction-stir welding of 5083 marine-grade aluminum alloy. In: Proceedings of IMechE Part B: Journal of Engineering Manufacture, p 8
30. Abbas M, Mehani N, Mittal A (2014) Feasibility of underwater friction stir welding and its optimization using taguchi method. Int J Eng Sci Res Technol 3(7):700–710
31. Zhang H, Liu H (2012) Characteristics and formation mechanisms of welding defects in underwater friction stir welded aluminum alloy. Metallogr Microstruct Anal 269–281

32. Mofid MA, Abdollah-zadeh A, Malek Ghaini F (2012) The effect of water cooling during dissimilar friction stir welding of Al alloy to Mg alloy. Mater Des 36:161–167
33. Mofid MA, Abdollah-Zadeh A, Ghaini FM, Gu CH (2012) Submerged Friction-Stir Welding (SFSW) underwater and under liquid nitrogen: an improved method to join Al alloys to Mg alloys. Miner Met Mater Soc ASM, Int 2012(43A):5106–5114
34. Majumdar JD (2006) Underwater welding—present status and future scope. J Naval Architect Mar Eng 3:39–48

Three Response Optimization of Spot-Welded Joint Using Taguchi Design and Response Surface Methodology Techniques

F. A. Ghazali, Z. Salleh, Yupiter H. P. Manurung, Y. M. Taib, Koay Mei Hyie, M. A. Ahamat and S. H. Ahmad Hamidi

Abstract One of the main challenges in correlating welding parameters and weld quality is its complexity to include as many as possible factors. In this research, the effects of spot welding parameters on weld quality were investigated. The effects of weld time, weld current, and electrode force on the sizes of fusion zone and heat affected zone, and tensile-shear load were studied. These welding parameters and weld quality were analysed using the three response Taguchi L_9 orthogonal array method in Minitab 17. Second-order regression models of fusion zone size, heat affected zone size and tensile-shear load were constructed by adapting Response Surface Method. The optimum weld time was 0.2 s, weld current of 10 kA and the required electrode force was 2.3 kN. These parameters were within 5% discrepancies with the experiment results. Weld current was the most important welding parameter that determines the weld quality, with the contribution of 69%. From our observation, the failure mode was the pullout type, a generally accepted failure for welded joint. The outcomes of this research contributed to the advancement in optimization technique for RSW joint, by increased the number of weld quality from two to three response.

Keywords Three response optimization · Taguchi method · Response surface methodology · Spot weld

---

F. A. Ghazali (✉) · M. A. Ahamat · S. H. Ahmad Hamidi
Universiti Kuala Lumpur, 1016, Jalan Sultan Ismail, 50250 Kuala Lumpur, Malaysia
e-mail: farizahadliza@unikl.edu.my

Z. Salleh · Y. H. P. Manurung · Y. M. Taib
Faculty of Mechanical Engineering, Universiti Teknologi MARA (UiTM), 40000 Shah Alam, Selangor, Malaysia

K. M. Hyie
Faculty of Mechanical Engineering, Universiti Teknologi MARA (Pulau Pinang), 13500 Permatang Pauh, Pulau Pinang, Malaysia

M. Awang (ed.), *The Advances in Joining Technology*, Lecture Notes in Mechanical Engineering, https://doi.org/10.1007/978-981-10-9041-7_7

1 Introduction

Resistance spot welding (RSW) is a rapid joining technique to join thin shell assemblies, such as components in automotive industry. One single typical car body consists about three hundred sheet metal parts, joined by thousands of spot-weld. The strength of these joints solely relies on the quality of its weld. Therefore, it is important to use appropriate setting of parameters in RSW. Those parameters are weld current, weld time, electrode force and holding time [1]. Use of correct values for these parameters is critical to obtain the optimal quality of weld joint. In brief, these parameters will affect the sizes of fusion zone (FZ) and heat affected zone (HAZ), mechanical properties and fatigue life of a welded joint. However, the correlations between the RSW parameters and weld quality are very scarce.

Low-carbon steel is one of the commonly used materials in various industries. Thus, it is essential to understand the behaviour of low-carbon steel that undergoes RSW joining technique. Numerous researches have been conducted on the RSW of low-carbon steel; but no attempt had been made to correlate the effect of RSW parameters to three welded joint quality. Earlier researches only studied the effect of one parameter (e.g. weld current) on the specific joint quality (e.g. size of fusion zone). By neglecting other RSW joint quality, the integrity of the welded joint is debateable. Since most of industrial applications involving RSW low-carbon steel requires a good quality welded joint, a study on three-response optimization become a necessity.

The effect of four RSW parameters, namely weld current, weld time, electrode force, and holding time on tensile-shear strength of welded joint was investigated [2]. Artificial neural network was adapted to investigate the correlation between these parameters and the tensile shear strength of the joint. The sizes of fusion zone and heat affected zone were not investigated. The effect of welding parameters on the tensile shear strength for the spot-weld joint of galvanized steel sheets was reported by Thakur et al. [3]. In their work, they used Analysis of Variance (ANOVA) to evaluate the level of importance for each welding parameters. Again, only one welded joint quality was considered in [3].

An improvement on the number of welded joint quality based on fusion Zone and heat affected zone was recently published [4]. The RSW parameters were optimised using the multiple-objective quality method. Another study reported the correlation between weld current, weld time and electrode force on tensile shear strength and fusion zone using Taguchi method to determine the optimum RSW parameters [5].

An optimal resistance spot welding parameters for a 1.0 mm thickness low carbon steel by considering multi-welded joint quality was reported in [6]. The setting of welding parameters was determined using L_9 Taguchi orthogonal array experimental design method, yielded the optimal processing conditions for spot welded of low carbon steel. The optimum welding parameter for two responses was obtained using multi-signal to noise ratio (MSNR) and the significant level of the welding parameters was further analysed using analysis of variance technique.

In this paper, the welding parameters were optimized based on three-response of welded joint (fusion zone, heat affected zone and tensile-shear load) by adapting Taguchi L_9 orthogonal array and central composite design method. RSW samples were prepared based on 29 parameters combinations. The response surface methodology was used to construct the second-order correlations between fusion zone, heat affected zone and tensile-shear load and welding parameters which were validated by comparison of predicted and experiment results. The mode of failure was evaluated to determine the integrity of these welded joints.

2 Methodology

This section presents the method on preparation of low-carbon steel RSW samples, measurement methods of the sizes of fusion and heat affected zones and the tensile-shear load test procedure. The Taguchi L_9 orthogonal array and Central Composite Design methods were used in the selection of values of welding parameters.

2.1 RSW Samples Preparation and Equipment

The low-carbon steel sheet with a thickness of 1.2 mm was cut into 105 mm × 45 mm piece, in accordance to the AWS 8.9 m standard. Then, two pieces of 105 mm × 45 mm × 1.2 mm sheet was welded using JPC 75 kVa spot welder machine in a single lap shear joint type, with the overlap of 35 mm. The machine was set to AC waveform, and the 5 mm diameter electrode was truncated at 30°. The chemical composition of the low-carbon steel is available in Table 1.

2.2 RSW Parameters Selection

The range of welding parameters as recommended in the manufacturer datasheet of JPC 75 kVa spot welder machine are available in Table 2. These data were used as the upper and lower limits of welding parameters.

For the Taguchi L_9 orthogonal array optimization method, the welding parameters and their factor levels are presented in Table 3. The values used for Level 2 were taken

Table 1 Chemical composition (%) of low carbon steel [7]

C	S	Mn	P	S	Fe
0.05	0.01	0.21	0.011	0.005	Bal.

Table 2 Range of welding parameters [8]

Variable	Range
Weld time (s)	0.17–0.31
Electrode force (kN)	2.0–3.4
Weld current (kA)	7.3–10.7

Table 3 Welding parameters and their factor levels for Taguchi L_9 orthogonal array

Symbol	Welding parameter	Unit	Level 1	Level 2	Level 3
WT	Weld time	s	0.2	0.24	0.28
EF	Electrode force	kN	2.3	2.7	3.1
WC	Weld current	kA	8.0	9.0	10.0

Table 4 Combinations of factor level in Taguchi L_9 orthogonal array

Experiment number	Factor level		
	WT	EF	WC
1	1	1	1
2	2	1	2
3	3	1	3
4	1	2	2
5	2	2	3
6	3	2	1
7	1	3	3
8	2	3	1
9	3	3	2

Table 5 Parameters and levels applied in CCD for spot weld of low carbon steel

Symbol	Parameter	Unit	Level				
			$-\alpha$	−1	0	+1	$+\alpha$
WT	Weld time	s	0.17	0.20	0.24	0.28	0.31
EF	Electrode force	kN	2.0	2.3	2.7	3.1	3.4
WC	Weld current	kA	7.3	8	9	10	10.7

from the average of maximum and minimum values for each variables in Table 2. In Table 4, nine combinations of factor level for welding parameters are specified.

The Central Composite Design (CCD) method was adapted to construct the second-order model to correlate welding parameters and welded joint quality. Five levels were used in CCD, and the values of parameters were tabulated in Table 5. A total of 20 parameters combinations were tested through experiments (Table 8).

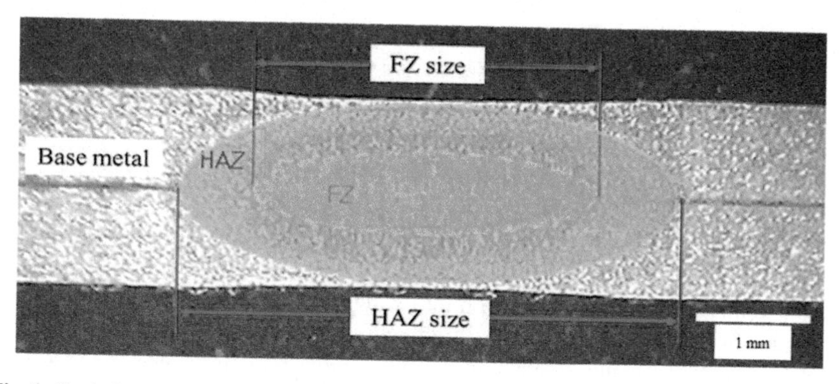

Fig. 1 Typical macrograph of weld zones

2.3 *Measurement of FZ and HAZ Sizes*

The welded sample was cut transversely at its center line using a cutter blade to expose its fusion zone and heat affected zone. Then, the surface was etched using 2% Nital solution for 5 s at room temperature. The macrograph of weld zone was captured using a zoom stereo microscope (Olympus BX 41 M Microscope) equipped with an image analysis system (IMAPS Version 4.0 Professional Edition Software). The typical shape of macrograph of weld zone, fusion zone, and heat affected zone is shown in Fig. 1.

2.4 *Evaluation of Tensile-Shear Load*

For tensile-shear load test, the tests were conducted according to the ISO 14323 standard. An Instron 200 kN universal testing machine was used to execute the tensile-shear load test. The crosshead rate was set at 2 mm/min. Three samples were prepared for each combination of RSW parameters, and the average tensile-shear load was taken for comparison with the calculated value from the second-order model that was constructed from CCD experiments results. After the samples failed, its failure modes were classified into interfacial and nugget pullout by visual inspection.

3 Results and Discussion

In this section, the findings on Taguchi L_9 OA and CCD optimization methods, and the failure mode of RSW sample are presented.

3.1 Three-Response Taguchi

The results of the experiments using Taguchi L_9 orthogonal array for FZ size, HAZ size and tensile-shear load is depicted in Table 6. The size of FZ was between 3.05 and 3.50 mm while the HAZ size was between 4.09 and 4.34 mm. For the tensile-shear load, the highest was 8.82 kN.

The results of analysis of variance (ANOVA) for the welding responses in Table 6 are presented in Table 7. ANOVA technique enables the evaluation of significant factors in three-response optimization for FZ size, HAZ size, and tensile-shear load. The value of F ranks the magnitude of influence of welding parameters on the weld-joint quality. A parameter with a higher value of F has more influence on the welded joint quality [9]. In our work, weld current is the most significant welding parameters affecting the welded joint quality whereas weld time has the lowest significance.

A further analysis confirmed that the most important factor is the weld current with 68.68%, a 44% more than the electrode force. The weld time has the lowest contribution of 6.43%. In short, a higher percentage of contribution means the parameter has more influence on the welded joint quality. Our results agreed with [6, 10], where weld current is the most important parameter in RSW. It was estimated that the

Table 6 Weld quality for Taguchi L_9 orthogonal array

Experiment number	FZ size (mm)	HAZ size (mm)	Tensile-shear load (kN)
1	3.05	4.14	7.89
2	3.06	4.41	7.84
3	3.06	4.19	8.50
4	3.32	4.28	8.43
5	3.34	4.34	8.55
6	3.36	4.32	8.53
7	3.36	4.26	8.48
8	3.41	4.19	8.48
9	3.50	4.09	8.82

Table 7 Results of ANOVA for the significance of welding parameters

Welding parameter	DF	Sum of squares	Mean of squares	F	P	Contribution
Weld time	2	4.851	2.4255	12.59	0.006	6.43
Weld current	2	51.811	25.9055	134.45	0.042	68.68
Electrode force	2	18.876	9.388	48.72	0.011	24.89
Error	2	0.146	0.073			
Total	8	75.854				100

variance for these experiments was 0.146, an indicator of an acceptable experiment design.

From the statistical point of view, the variables that have *p*-value that is less than 0.05 is considered statistically significant to the welded joint quality. The analysis showed that weld time, weld current and electrode force were all significant in determining the welded joint quality. All of these weld parameters were considered in the developed response surface model reported in Sect. 3.2.

3.2 Response Surface Methodology

The second order response surface model for the significant parameters affecting the FZ size, HAZ size and tensile-shear load was constructed using the three responses Taguchi L_9 orthogonal array method. The CCD (Central Composite design) technique was used to generate 20 combinations of welding parameters (Table 8). The combinations were based on 8 factorial points (standard order 1–8), six star points (standard order 9–14) and six replicates of the center point (standard order 15–20).

The design was expanded to evaluate several points which increased the chance in detecting the response at which the optimum parameters combinations occurred. The center point represented a set of experiment conditions at which six independent replicates were run. The variation between those conditions reflected the variability of all designs. It was used to estimate the standard deviation. All optimization experiments were conducted randomly in one block of measurement.

A second-order model for FZ, HAZ and tensile shear load was constructed to describe the behavior of each response. The second-order models for the FZ size, HAZ size and tensile-shear load in terms of un-coded variables with all significant terms are given in Eqs. (1)–(3), respectively:

$$\begin{aligned} \mathrm{FZ} = {} & 3.61 + 0.15\mathrm{WT} + 0.08\mathrm{EF} + 0.15\mathrm{WC} - 0.13\mathrm{WT}^2 \\ & + 0.03\mathrm{EF}^2 - 0.04\mathrm{WT} * \mathrm{EF} - 0.07\mathrm{WT} * \mathrm{WC} + 0.02\mathrm{EF} * \mathrm{WC} \end{aligned} \quad (1)$$

$$\begin{aligned} \mathrm{HAZ} = {} & 4.68 + 0.12\mathrm{WT} + 0.07\mathrm{EF} + 0.05\mathrm{WC} - 0.06\mathrm{WT}^2 + 0.03\mathrm{EF}^2 \\ & - 0.02\mathrm{WC}^2 - 0.06\mathrm{WT} * \mathrm{EF} - 0.04\mathrm{WT} * \mathrm{WC} + 0.07\mathrm{EF} * \mathrm{WC} \end{aligned} \quad (2)$$

$$\begin{aligned} \text{Tensile-shear load} = {} & 8.62 + 0.45\mathrm{WT} + 0.28\mathrm{EF} \\ & + 0.23\mathrm{WC} - 0.22\mathrm{WT}^2 - 0.03\mathrm{EF}^2 - 0.03\mathrm{WC}^2 \\ & - 0.20\mathrm{WT} * \mathrm{EF} - 0.27\mathrm{WT} * \mathrm{WC} + 0.11\mathrm{EF} * \mathrm{WC} \end{aligned} \quad (3)$$

The analysis showed that the regression model and each variable term (linear, square and interaction) in the model had a *p*-value that was less than 0.05. To test the global fit of the model, the coefficient of determination (R^2) was evaluated. The R^2 for tensile-shear load was 0.997, denoting that the sample variation of 99.7% is attributed to the regressors in the model and only 0.3% of the total variability is

Table 8 Design matrix of CCD and experimental design

Run order	Parameter			Response		
	Weld time (s)	Weld current (kA)	Electrode force (kN)	FZ size (mm)	HAZ size (mm)	Tensile-shear load (kN)
1	0.24	9	2.7	3.61	4.68	8.63
2	0.24	9	2.7	3.60	4.69	8.67
3	0.20	8	2.3	3.02	4.32	7.01
4	0.28	10	2.3	3.74	4.65	8.50
5	0.24	10.7	2.7	3.81	4.70	8.96
6	0.20	10	3.1	3.79	4.80	8.96
7	0.28	8	2.3	3.40	4.77	8.88
8	0.24	9	2.0	3.58	4.68	8.13
9	0.24	9	2.7	3.62	4.66	8.59
10	0.20	8	3.1	3.11	4.50	7.80
11	0.24	7.3	2.7	3.48	4.56	8.15
12	0.24	9	2.7	3.62	4.66	8.58
13	0.17	9	2.7	3.05	4.34	7.24
14	0.20	10	2.3	3.29	4.40	7.80
15	0.24	9	3.4	3.74	4.84	8.98
16	0.31	9	2.7	3.50	4.71	8.78
17	0.28	8	3.1	3.66	4.68	8.80
18	0.24	9	2.7	3.61	4.69	8.60
19	0.24	9	2.7	3.60	4.68	8.65
20	0.28	10	3.1	3.72	4.88	8.96

not explained by the model. The R^2 for the FZ size and HAZ size were 90.71 and 97.78%, respectively.

3.3 *Confirmation Tests*

The results obtained from Eqs. (1) to (3) were compared with the experiments (Table 9). The percentage of discrepancies between confirmation experiments and prediction (Eqs. 1–3) were less than 3.7%. From the ANOVA analysis, the optimum welding parameters are as follow: weld time of 0.2 s, electrode force at 2.3 kN and a weld current of 10 kA.

The investigation on failure mode was carried out using tensile-shear test until the welded joint was failed. The failure was classified either as the well button pullout or interfacial fracture. For the RSW joint of the samples, it was observed that the failure mode was a pullout failure as in Fig. 2. This type of failure is generally the preferred

Table 9 Results of the confirmation tests for FZ size, HAZ size and tensile-shear load

Parameter	Value from Eqs. (1) to (3)	Experiment				Discrepancy (%)
		1	2	3	Average	
FZ size (mm)	3.50	3.49	3.41	3.43	3.44	1.71
HAZ size (mm)	4.54	4.63	4.55	4.58	4.59	1.09
Tensile-shear load (kN)	8.52	8.02	8.13	8.47	8.21	3.64

failure mode in resistance spot welding, due to its higher plastic deformation and energy absorption characteristics [11–13].

For a welded joint, the failure usually occurs at the softer base metal. The samples tested in this work produced the complete button pullout. In the pullout mode, failure commonly occurs via withdrawal of fusion zone (Fig. 2a) from one of the sheets [14, 15]. One of the signs of insufficient mechanical strength in RSW joint is if the failure occurs in the mode of interfacial failure or through the fusion zone failure [16].

In Fig. 2b, the failure of the spot weld appears to be initiated approximately at the middle of the fusion zone circumference before it was propagated through necking/shear along the fusion zone circumference. Once the welded joint lost its strength, the upper sheet was torn off. This observed failure mechanism is in agreement with other literatures [15, 17]. It can be concluded that the strength of weld joint produced by the combination of optimized parameters were able to create a joint with an acceptable failure mode.

4 Conclusions

An experimental design using a three-response optimization using the Taguchi L_9 orthogonal array method was used to determine the effects of welding parameters (weld time, weld current and electrode force) on the fusion zone size, heat affected zone size and tensile-shear load, simultaneously. Three second-order models for fusion zone size, heat affected zone size and tensile-shear load were constructed and validated by comparing the predicted values and experiment results. These second-order models were satisfactory since the values of R^2 for the global fitting of fusion zone size, heat affected zone size, and tensile-shear load were 0.907, 0.9778 and 0.994, respectively. It was found that the weld time of 0.2 s, electrode force at 2.3 kN and a weld current of 10 kA produced the desired welded joint quality. The experimental results obtained using these optimum welding parameters and the prediction based on the second-order models were within 3.7% discrepancies.

Fig. 2 Failure mode of tensile-shear test **a** pullout and **b** tearing

(a)

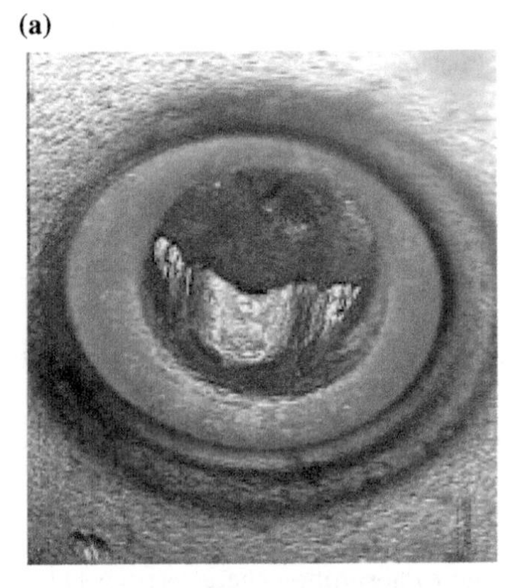

(b)

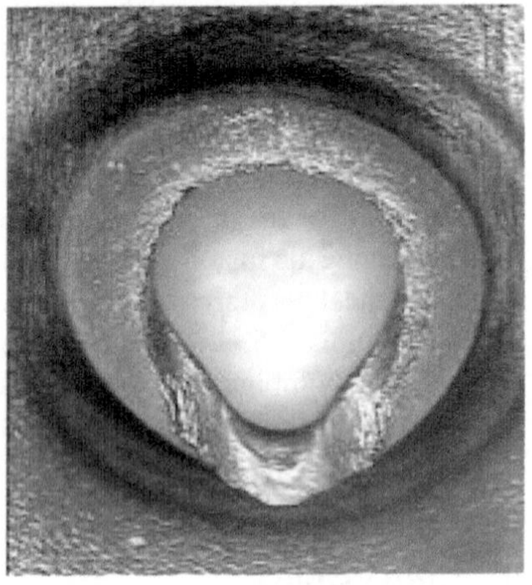

References

1. Muhammad N, Manurung YHP (2012) Design parameter selection and optimization of weld zone development in resistance spot welding. World Acad Sci Eng Technol 71
2. Hamidinejad SM, Kolahan AF, Kokabi AH (2012) The modeling and process analysis of resistance spot welding on galvanized steel sheets used in car body manufacturing. Mater Des 34:9
3. Thakur AG, Rao T, Mukhedkar MS, Nandedkar VM (2010) Application of Taguchi method for resistance spot welding of galvanized steel. ARPN J Eng Appl Sci 5(11):5
4. Muhammad N, Manurung YHP, Jaafar R, Abas SK, Tham G, Haruman E (2013) Model development for quality features of resistance spot welding using multi-objective Taguchi method and response surface methodology. J Intell Manuf 24:1175–1183

5. Pandey AK, Moeed KM, Khan MI (2013) Investigation of the effect of current on tensile strength and nugget diameter of spot welds made on AISI-1008 steel sheets. Int J Tech Res Appl 1:8
6. Muhammad N, Manurung YHP, Hafidzi M, Abas SK, Tham G, Rahim MRA (2012) A quality improvement approach for resistance spot welding using multi-objective Taguchi Method and response surface methodology. Int J Adv Sci Eng Inf Technol 2:6
7. Ghazali FA, Manurung YHP, Mohamed MA (2014) Multi-response optimization using Taguchi method of resistance spot welding parameters, Appl Mech Mater 660
8. Ghazali FA, Berhan MN, Manurung YHP, Salleh Z, Abdullah S (2015) Tri-objective optimization of carbon steel spot-welded joints. J Teknol 76(11)
9. Ahmadi H, Arab NBM, Ghasemi FA (2014) Optimization of process parameters for friction stir lap welding of carbon fibre reinforced thermoplastic composites by Taguchi method. J Mech Sci Technol 28(1):279–284
10. Esme U (2009) Application of Taguchi method for the optimization of resistance spot welding process. Arab J Sci Eng 34
11. Pouranvari M, Marashi SPH (2010) On the failure of low carbon steel resistance spot welds in quasi-static tensile–shear loading. Mater Des 31(8):3647–3652
12. Abadi MMH, Pouranvari M (2010) Correlation between macro/micro structure and mechanical properties of dissimilar resistance spot weld of AISI 304 Austenitic stainless steel and AISI 1008 low carbon steel. Assoc Metall Eng Serbia AMES 16(2):14
13. Pouranvari M, Ranjbarnoodeh E (2011) Resistance spot welding characteristic of ferrite-martensite DP600 dual phase advanced high strength steel-part II: failure mode. World Appl Sci J 15(11):1527–1531
14. Pouranvari M (2011) Effect of welding parameters on the peak load and energy absorption of low-carbon steel resistance spot welds. Int Sch. Res Netw ISRN Mech Eng 2011:1–7
15. Marashi P, Pouranvari M, Amirabdollahian S, Abedi A, Goodarzi M (2008) Microstructure and failure behavior of dissimilar resistance spot welds between low carbon galvanized and austenitic stainless steels. Mater Sci Eng, A 480:5
16. Jahandideh A, Hamedi M, Mansourzadeh S, Rahi A (2011) An experimental study on effects of post-heating parameters on resistance spot welding of SAPH440 steel. Sci Technol Weld Join 16(8):669–675
17. Pouranvari M, Marashi P (2009) Failure behaviour of resistance spot welded low. Assoc Metall Eng Serbia AMES 15(3):8

A Review on Mechanical Properties of SnAgCu/Cu Joint Using Laser Soldering

Nabila Tamar Jaya, Siti Rabiatull Aisha Idris and Mahadzir Ishak

Abstract This paper focuses on publishing the review of mechanical properties at the interface between SnAgCu (SAC) solder and Copper substrate after laser soldering. This involved on basic principles of solder alloy and Copper diffusion mechanism. In addition, this paper also reviews laser solder effects towards mechanical properties of the solder joint. This paper approach on the review of the solder joint strength which regards to intermetallic compound type and thickness. The output of this paper is to create an understanding for the readers about variance of laser soldering parameters and its effect on the mechanical properties of the solder joint by including discussion from other research paper findings.

Keywords Mechanical properties · Soldering · Metallurgy · Intermetallic compound

1 Introduction

Solder alloy is a filler material that is used in soldering for joining two metals part together where their liquidious temperature is below 450 °C [1–3]. Some of the characteristic of a solder alloy that need to be noted on are their materials, microstructure, the changes at interface during metallization phase, good wetting and manufacturability (have low melting and short melting-temperature range) as all of these factors will contribute to a quality solder joint reliability [4] Meanwhile, the diffusion between solder and copper will form intermetallic at the interface which then develop an adequate wetting properties of most lead-free and Sn-based solder that is important for solderability of a joint [5, 6]. The common phases that are formed in the intermetallic layer for copper (Cu)/tin (Sn) diffusion system are Cu_3Sn (at interface with copper)

N. T. Jaya · S. R. A. Idris (✉) · M. Ishak
Faculty of Mechanical Engineering, University Malaysia Pahang, 26600 Pekan, Pahang, Malaysia
e-mail: rabiatull@ump.edu.my

M. Awang (ed.), *The Advances in Joining Technology*, Lecture Notes in Mechanical Engineering, https://doi.org/10.1007/978-981-10-9041-7_8

and Cu_6Sn_5 (at interface with the tin) and the thickness of this layer depends on the concentration of Cu that presents in the solder alloy [7, 8].

Due to the environmental and health concerns, Sn–Pb solders have been banned and the studies on lead-free solder have been done world widely in conjunction of supporting Waste Electrical and Electronic Equipment (WEEE) and Restriction of the Use of Hazardous Substances in Electrical and Electronic Equipment (RoHS Directive) [9–11]. Lead-free solder such as SnCu and SnAgCu (SAC) solder alloys are seen as great potential substitute of Sn–Pb solder alloys due to its good wettability, high electrical conductivity, good mechanical properties and economically costing [12–15]. Furthermore, Cu has high availability, good electrical and thermal conductivity, therefore it has always been chosen for any metallization layers medium in the manufacturing of electrical components [16–18]. SAC solders have good wetting property, high creep resistance and when they are in molten form, these solders have high reaction rate with metals [19, 20].

Laser soldering technology is known to have been used since many years before due to its highly advanced selective soldering whereas the heat is applied only on the desired spot precisely, mechanically controlled by the machine and thus result in high production rate [10, 17, 21–23]. Meanwhile, as to compare with conventional method which is reflow soldering, the laser soldering has contactless temperature measurement during laser process, rapid rise and fall temperature cycles that will certainly minimize the thermal damage especially for heat-sensitive components [19, 24, 25]. It has also been claimed that soldering by using laser technology produces a good electrical conductivity in the integrated circuits assembly and void defects are minimized due to its fast evaporation of flux [9, 26]. Therefore, laser soldering technology has been quite well-known of its advantages in electronic packaging industries.

Meanwhile, intermetallic compound formation (IMC) is an interfacial layer that form through a chemical reaction during joining process between the molten solder and its substrate which is usual the base metal [27, 28]. IMC is said to be brittle, thus an excessive of this layer could deteriorate a solder joint as its formation already alter both the composition and mechanical performance of the joint [28, 29]. However, in laser soldering technology, it has rapid heating and fast cooling rate, the intermetallic compound (IMC) formation is minimize where it is known that the brittleness of the IMC could contribute to a joint failure [22, 30, 31]. As compared to reflow soldering, the integrated circuits assembly will be exposed to heat more longer in time which then results in the IMC crystals to keep on growing, thus create a thick IMC layer and reduce the quality of the solder joint [32–34].

2 Process of Laser Joining

A laser usually needs three elements to operate which are, firstly an active medium (laser rod) with selectively populated level of energy. Secondly, a pumping process to create population inversion between the energy levels, and lastly is the resonant

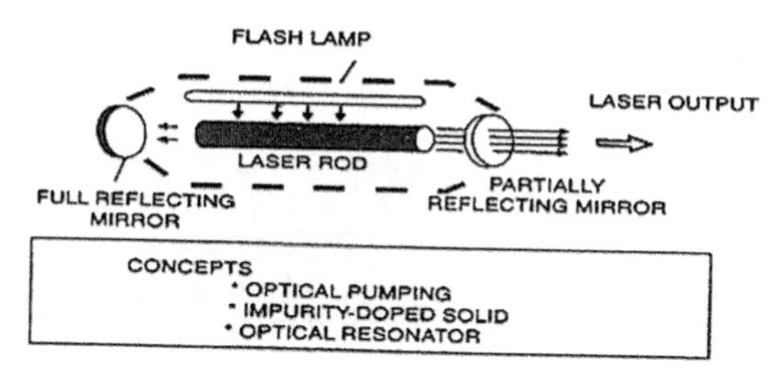

Fig. 1 The three main elements that involve in designing a laser [35]

Table 1 The comparison of different between laser sources for laser soldering [38]

	Diode laser	Fiber laser	Nd:YAG laser	CO_2 laser
Dimension (ratio)	1	1	100	1000
Electro-optical efficiency	40–50%	30%	3–17%	10–15%
Emission wavelength	800–1000 nm	1030–1090 nm	1064 nm	10,600 nm
Lifetime	>100,000 h	>100,000 h	>1000 h	~10,000 h
Maintenance	Low maintenance	Low maintenance	200–1000 h	Every 500 h
Investment/power	10–50 W	15–30 W	50–100 W	10–50 W
Beam guiding (fiber optics)	Yes	Yes	Yes	No

cavity which consist of active medium that serves as storage of the emitted radiation and creates feedback in order to retain the radiation coherence (Fig. 1) [35].

Fundamentally, lasers can be classified into two types which are continuous wave (CW) laser and pulse beam laser. In CW laser, the light was emitted in a steady continuous beam with low intensity. Gas laser was said to be categorised in this type. Meanwhile, for pulse beam laser, it emits powerful bursts of light with very short duration of period and the common lasers involved are crystals, glass and liquid types of lasers [35]. Flash lamp pump is used in pulsed beam lasers, while electric arc lamp is used for CW lasers [36].

Four main types of lasers have been found suitable to be used for laser soldering are semiconductor laser (diode laser), fiber laser, Nd:YAG laser (a solid state laser) and carbon dioxide laser (gas laser) [36–39]. Some characteristics of these four different type of lasers are described in Table 1 [38]. However, in this paper, the main focus will be on diode laser and fibre laser only due to its low maintenance and ease in operation which regards to its application in electronics industry.

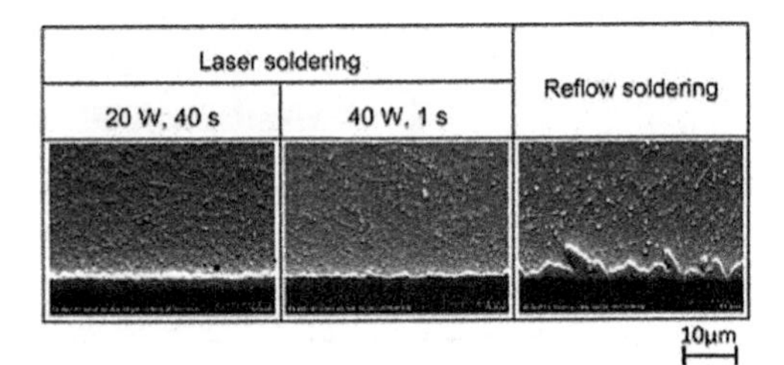

Fig. 2 Scanning electron microscopy (SEM) micrographs of the interface between the solder and Cu pad just after and reflow soldering [25]

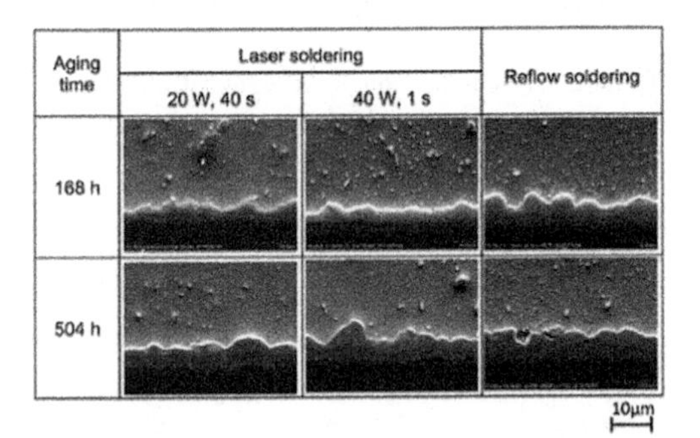

Fig. 3 Scanning electron microscopy (SEM) micrographs of the interface between the solder and Cu pad after isothermal aging for laser and reflow soldering [25]

2.1 Mechanical Properties of Solder Joint When Exposed to Laser Joining

In the study of Nishikawa [25], the formation and growth of the IMC layer at the Sn–Ag–Cu solder/Cu interface after laser soldering and during isothermal aging were investigated to elucidate the basic characteristics of the laser soldering process. The impact reliability of the joint soldered using laser soldering was also investigated. According to his work, it was found that the growth of the IMC layer during isothermal aging appears to be faster in the case of laser soldering than in the case of reflow soldering (Fig. 2). In the case of the reflow soldering, the IMC layer at the Sn–Ag–Cu/Cu pad interface has a typical scallop-like morphology. In contrast, a relatively thin IMC layer is formed at the interface in the case of the laser soldering, regardless of the laser irradiation condition. The thickness of this IMC layer was less than 1 μm. This is because the duration for which the solder was heated to a temperature greater than its melting temperature was much shorter during laser soldering than for reflow soldering. Hence, the use of laser soldering causes the reaction time between the molten solder and Cu pad to be very brief. After both samples undergoes the isothermal ageing for 168 h and 504 h, small voids are observed to appear more on the laser compared to the reflow soldering at the interface (Fig. 3). For both laser and reflow soldering, total IMC layer thickness increases with the increase of aging time.

The thickness is proportional to the square root of aging time, Fig. 4. The thickness of IMC usually is much thinner in laser due to its shorter soldering time [26]. It is

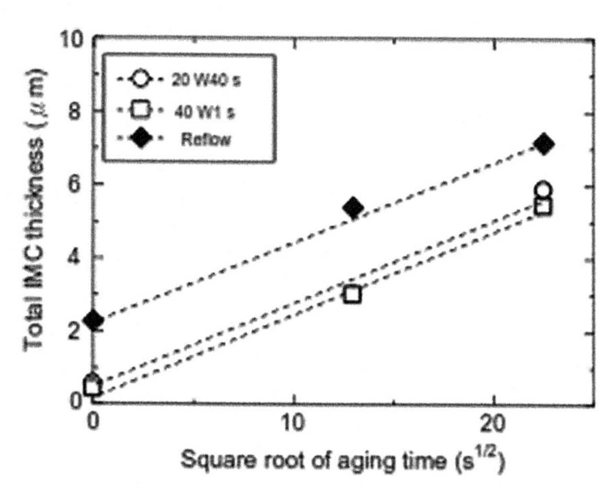

Fig. 4 Effect of aging time at 423 K on the total intermetallic compound (IMC) thickness at the solder/Cu pad interface [25]

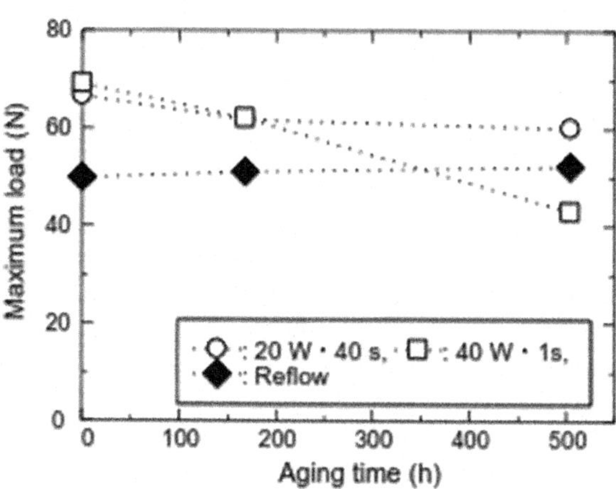

Fig. 5 Effect of aging time on the maximum load of the load-displacement curve obtained from the microimpact tester [25]

well known that voids in the IMC layer provide initiation sites and propagation paths for cracks whose formation degrades the reliability of solder joints as reported by Zeng et al. [14]. While Fig. 5 demonstrates the effect of aging time on the maximum load of the load-displacement curve obtained from the microimpact tester. In the as-soldered condition, the maximum load of joints soldered by the laser process was superior to that soldered by the reflow process because of the extremely thin IMC layer at the interface and the high hardness of the solder matrix.

In the study of Lui [40], they used laser to solder rapidly-solidified lead-free solders for two reasons, one is the good characteristics of laser soldering such as rapid melting, rapid cooling and small heat affected zone and the other is that the rapidly-solidified lead-free solders are not suitable for long time. Figure 6 shows the thickness of IMCs at rapidly solidified SAC305 interface under different scanning speed when the laser output power is 50 W. As can be seen the thickness of IMC becomes thinner with the scanning speed increases. When the scanning speed is less

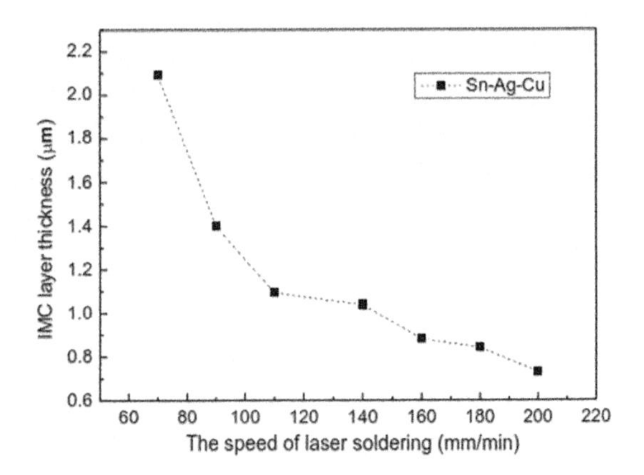

Fig. 6 The relationship of thickness of IMC and laser scanning speed, when the laser output power is 50 W [40]

than 50 mm/min, the IMC layer is too thick. According to the study of Yang et al., Cu dissolves from substrates into the molten solder during soldering in an amount that is related to the soldering process and after the soldering is complete, the dissolved Cu will precipitate in the solder matrix and at interface [41]. This precipitation amount and distribution would greatly affect the joint reliability.

In Fig. 7 it can be seen both of the rapidly-solidified lead-free solders and the as-cast lead-free solders conform to the same rule that the IMC layer becomes thicker with the aging time increases. However, in the subsequent aging process which is from one day to three days, the IMCs formed at rapidly-solidified Sn3.5Ag0.7Cu/Cu interface became thinner than that formed at as-cast Sn3.5Ag0.7Cu/Cu interface under the same aging time [40].

In the study of Xue [22], they performed an experiment in order to investigate the advantages of laser soldering technology on its application in the lead-free environments. In their experiments, Sn–3Ag–0.5Cu lead-free solder on Au/Ni/Cu pad were carried out by means of diode-laser and IR reflow soldering methods respectively. In his work, it was found that tensile force of QFP micro-joints increases gradually with the increase of laser output power (Fig. 8). The maximum force of QFP micro-joints is gained when the laser output power increases to 34.1 W, which shows that the PCB is wetted adequately by molten Sn–Ag–Cu lead-free solder, and the best mechanical properties of micro-joints are gained at the same time.

Meanwhile, they also compared the mechanical properties between laser and infrared (IR) reflow soldering as it shown in Fig. 9. In Fig. 9, It is concluded that mechanical properties of QFP micro-joints soldered by laser soldering system are better than those soldered by IR reflow soldering methods.

Other than that, An et al. [42] has done a research work on the fiber-laser soldering machine as the heating resource to solder the SAC/Cu in order to investigate the effort of laser soldering parameters on the morphology and the size of the Cu_6Sn_5 grains. From the result, it was showed that that the Cu_6Sn_5 grains were in scallop type with smooth surface. According to the study Yang et al., in lead-free solders, Sn is the main element and usually reacts with Cu substrates to form two interposing

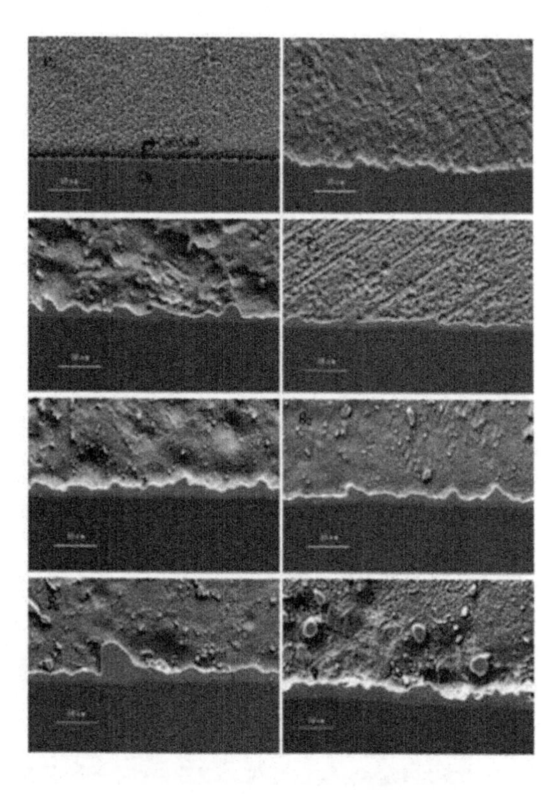

Fig. 7 The comparison of cross-sections of IMCs at rapidly solidified Sn3.5Ag0.7Cu/Cu and as cast Sn3.5Ag0.7Cu/Cu aging at 150 for (e) 0 day, (f1, f2) 1 day, (g1, g2) 2 days, (h1, h2) 3 days [40]

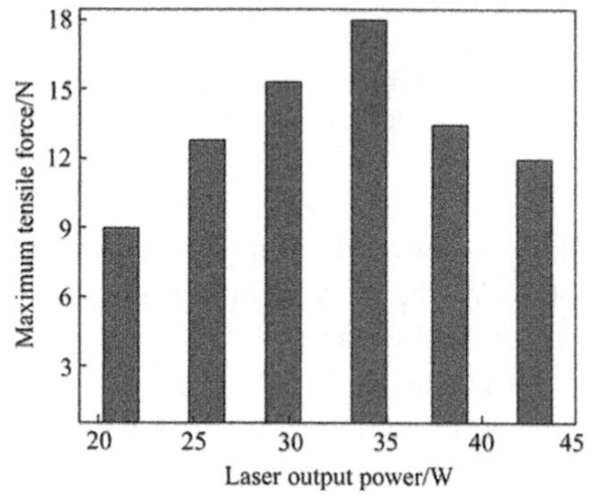

Fig. 8 The relationship between laser output power and maximum tensile force of QFP SAC micro-joints [22]

layers of Cu–Sn intermetallic compounds (IMCs) such as Cu_3Sn and Cu_6Sn_5 [41]. Meanwhile, as reported by Shang et al., the Cu_3Sn is formed next to the Cu, whereas the Cu_6Sn_5 is formed immediately adjacent to the solder [43]. In Fig. 10a, the grain of Cu_6Sn_5 is in scallop type with smooth surface. As seen in Fig. 10b, when the laser soldering output power was 40 W, the morphology of the Cu_6Sn_5 was prism-type. The size of the prism-type grains differed a lot, the length of the longest of the grains can be 8.24 μm while that of the shortest one was only 3.29 μm. In Fig. 10c shows

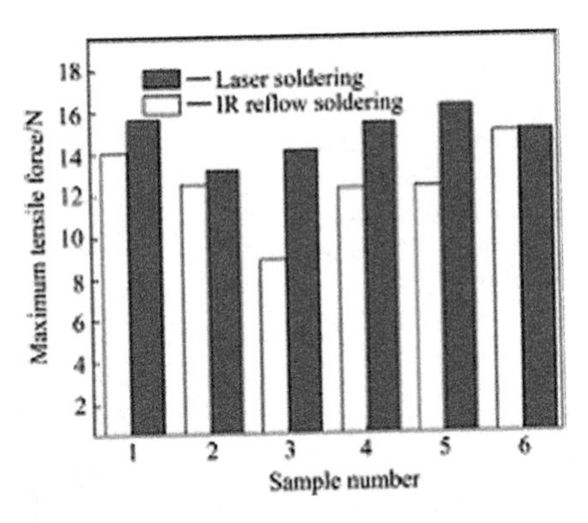

Fig. 9 The tensile properties of QFP micro-joints soldered [22]

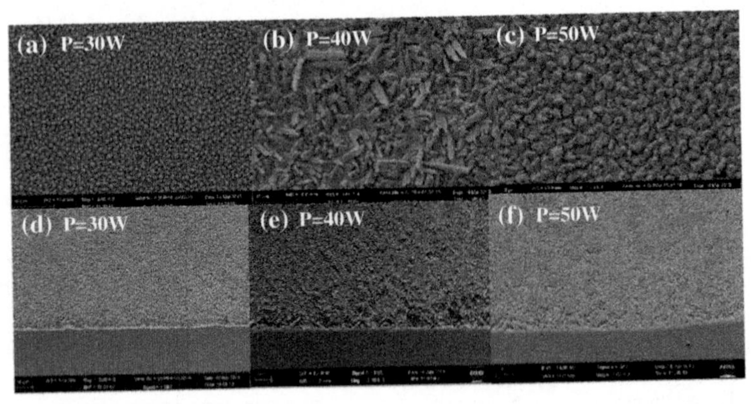

Fig. 10 Morphology of Cu_6Sn formed at SAC/Cu solder joint at laser scanning speed of 100 mm/min with different laser soldering output power: **a**, **b**, **c** Top view of Cu_6Sn and **d**, **e**, **f** Cross-sectional view of Cu_6Sn [42]

the top-view of the Cu_6Sn_5 grains when soldered at the laser soldering output power of 50 W, the morphology of the Cu_6Sn_5 grains was bottom closely arranged with smooth top surface.

When the laser soldering scanning speed was 230 mm/min, the Cu was in scallop-type grains with faceted surface Fig. 11a. The size of the grains was unified and fine but when the scanning speed was reduced to 180 mm/min, many more grains with faceted surface can be found which indicated the Cu_6Sn_5 grains had the tendency to convert from scallop type to prism type Fig. 11b. Meanwhile, the size of the grains was bigger. As shown in Fig. 11c, when the laser soldering scanning speed was 140 mm/min, nearly all the Cu_6Sn_5 grains were in prism type. There was an obvious difference between the prism type grains. Some of the grains were small prism while some grains grew into long lengths.

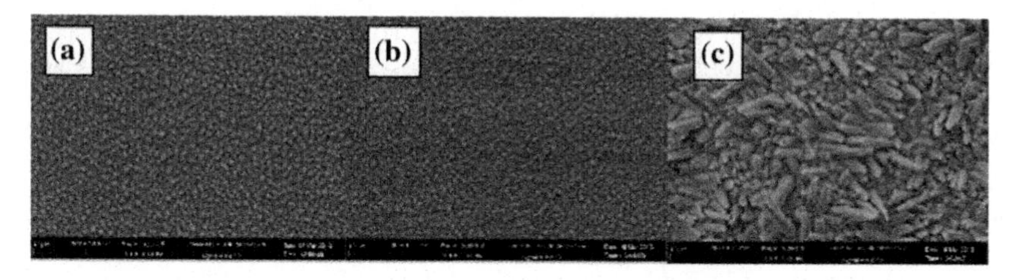

Fig. 11 Top-view of Cu_6Sn formed at SAC/Cu solder joint at laser output power of 50 W with different laser soldering scanning speed: **a** 230 mm/min, **b** 180 mm/min and **c** 140 mm/min [42]

3 Conclusions

As a conclusion, laser soldering managed to retard the IMC formation and increase the strength of the solder joint. The thickness of the IMC by laser soldering reportedly to be less than 1 μm. The thickness of IMC also is claimed become thinner as the power increased, but at some point, if the power is too high, the formation of IMC starts to discontinue as stated by many studies, the IMC layer must be thick enough to obtain very strong solder joint, but it must be kept to a minimum for very good electrical and also thermal performance. At the same time, it can be concluded that it is thick enough for optimal mechanical characteristics (as proven by the mechanical tests). It is known, a too thick layer of IMC could deteriorate the mechanical strength as the layer is brittle and porous. However, some studies also shown that with the increasing of aging time, mostly SAC solder group solder joint strength started to decrease. Cracks are said to initiate at the interface between solder and IMC layer or between IMC and IMC layer itself. Soldering by using laser method also favored maybe due to its time less consuming, energy saving because the power applied punctually or localized at one point at one time. The electrical and mechanical tests proved a good solder joint performance.

References

1. Hwang JS (2012) Solder paste in electronics packaging: technology and applications in surface mount, hybrid circuits, and component assembly. Springer, Netherlands
2. Silvera J (2012) Soldering made simple: easy techniques for the kitchen-table jeweler. Kalmbach Publishing Company, Books Division, Waukesha
3. Rao (2011) Manufacturing technology: foundry, forming and welding,2e. McGraw-Hill, New York
4. Frear DR, Burchett SN, Morgan HS, Lau JH (1994) Mechanics of solder alloy interconnects. Springer, US
5. Gnecco F, Ricci E, Amore S, Giuranno D, Borzone G, Zanicchi G, Novakovic R (2007) Wetting behaviour and reactivity of lead free Au–In–Sn and Bi–In–Sn alloys on copper substrates. Int J Adhes Adhes 27(5):409–416
6. Novakovic R, Lanata T, Delsante S, Borzone G (2012) Interfacial reactions in the Sb–Sn/(Cu, Ni) systems: wetting experiments. Mater Chem Phys 137(2):458–465

7. Yang ZJ, Yang SM, Yu HS, Kang SJ, Song JH, Kim KJ (2014) IMC and creep behavior in lead-free solder joints of Sn–Ag and Sn–Ag–Cu alloy system by SP method. Int J Automot Technol 15(7):1137–1142
8. Yu DQ, Wu CML, Law CMT, Wang L, Lai JKL (2005) Intermetallic compounds growth between Sn–3.5Ag lead-free solder and Cu substrate by dipping method. J Alloy Compd 392(1):192–199
9. Kim J-O, Jung J-P, Lee J-H, Jeong S, Kang H-S (2009) Effects of laser parameters on the characteristics of a Sn–3.5 wt%Ag solder joint. Met Mater Int 15(1):119–123
10. Illyefalvi-Vitez Z, Balogh B, Baranyay Z, Farmer G, Harvey T, Kirkpatrik D, Kotora G, Ruzsics N (2007) Laser soldering for lead-free assembly. In: ISSE 2007—30th international spring seminar on electronics technology 2007; conference proceedings: emerging technologies for electronics packaging, pp 471–476
11. Wang M, Wang J, Ke W (2017) Corrosion behavior of Sn–3.0Ag–0.5Cu lead-free solder joints. Microelectron Reliab 73:69–75
12. Abd El-Rehim AF, Zahran HY (2017) Investigation of microstructure and mechanical properties of Sn–xCu solder alloys. J Alloy Compd 695:3666–3673
13. Zeng G, Xue S, Zhang L, Gao L, Lai Z, Luo J (2010) Properties and microstructure of Sn–0.7Cu–0.05Ni solder bearing rare earth element Pr. J Mater Sci: Mater Electron 22(8):1101–1108
14. Zeng G, Xue S, Zhang L, Gao L (2011) Recent advances on Sn–Cu solders with alloying elements: review. J Mater Sci: Mater Electron 22(6):565–578
15. Cheng S, Huang C-M, Pecht M (2017) A review of lead-free solders for electronics applications. Microelectron Reliab 75:77–95
16. Yao P, Li X, Liang X, Yu B (2017) Investigation of soldering process and interfacial microstructure evolution for the formation of full Cu3Sn joints in electronic packaging. Mater Sci Semicond Process 58(Nov 2016):39–50
17. Kibushi R, Hatakyeama T, Imai D, Nakagawa S, Ishizuka M (2013) Optimal laser condition for laser soldering in cream and ring solder. In: 2013 IEEE 3rd CPMT symposium Japan (ICSJ 2013)
18. Ervina EMN, Aisyah MA (2012) A review of solder evolution in electronic application. Int J Eng Appl Sci 1(1):1–10
19. Nishikawa H, Iwata N, Takemoto T (2011) Effect of heating method on microstructure of Sn–3.0Ag–0.5Cu solder on Cu substrate
20. Hamada N, Uesugi T, Takigawa Y, Higashi K (2012) Effects of Zn addition and aging treatment on tensile properties of Sn–Ag–Cu alloys. J Alloys Compd 527(Supplement C):226–232
21. Nicolics J (1992) Optimization of process parameters for laser soldering of surface mounted devices 15(6):1155–1159
22. Han Z-J, Xue S-B, Wang J-X, Xin Z (2008) Mechanical properties of QFP micro-joints soldered with lead-free solders using diode laser soldering technology. Trans Nonferrous Met Soc China, 2–6
23. Wang J-X, Xue S-B, Fang D-S, Ju J-L, Han Z-J, Yao L-H (2006) Effect of diode-laser parameters on shear force of micro-joints soldered with Sn–Ag–Cu lead-free solder on Au/Ni/Cu pad. Trans Nonferrous Met Soc China 16:1374–1378
24. Liu W, Wang CQ, Tian YH (2008) Effect of laser input energy on AuSn X intermetallic compounds formation in solder joints with different thickness of Au surface finish on pads. Acta Metall Sinica (English Letters) 21(3):183–190
25. Nishikawa H, Iwata N (2015) Formation and growth of intermetallic compound layers at the interface during laser soldering using Sn–Ag Cu solder on a Cu pad. J Mater Process Technol 215(Supplement C):6–11
26. Bunea R, Svasta P (2011) Optimizing laser soldering of SMD components: from theory to practice, pp 55–58
27. Lu D, Wong CP (2016) Materials for advanced packaging. Springer International Publishing, Cham
28. Hare E (2013) Intermetallics in solder joints. http://www.semlab.com/papers, SEMLab Inc

29. Kang TY, Xiu YY, Hui L, Wang JJ, Tong WP, Liu CZ (2011) Effect of bismuth on intermetallic compound growth in lead free solder/cu microelectronic interconnect. J Mater Sci Technol 27(8):741–745
30. Flanagan A, Conneely A, Glynn TJ, Lowe G (1996) materials processing technology fine pitch electronic components. J Mater Process Technol 0136:531–541
31. Chung DDL (1995) Materials for electronic packaging. Elsevier Science, New York
32. Liu P, Yao P, Liu J (2009) Effects of multiple reflows on interfacial reaction and shear strength of SnAgCu and SnPb solder joints with different PCB surface finishes. J Alloy Compd 470(1):188–194
33. Noh BI, Koo JM, KIm JW, Kim DG, Nam JD, Joo J, Jung SB (2006) Effects of number of reflows on the mechanical and electrical properties of BGA package. Intermetallics 14(10):1375–1378
34. Ma HR, Wang YP, Chen J, Ma HT, Zhao N (2017) The effect of reflow temperature on IMC growth in Sn/Cu and Sn0.7Cu/Cu solder bumps during multiple reflows, pp 1402–1405
35. Ion J (2005) Laser processing of engineering materials: principles, procedures and industrial applications. pp 335
36. Ogochukwu ES (2013) Laser soldering. In: Y. Mastai, Rijeka (ed) Materials Science - Advanced Topics. InTech, 2013, p Ch 15
37. Dawson JW (2012) Fiber lasers: technology, applications and associated laser safety. Article
38. Qin Y (2010) Micromanufacturing engineering and technology. Elsevier Science, New York
39. Majumdar JD, Manna I (2012) Laser-assisted fabrication of materials. Springer, Berlin
40. Liu J, Ma H, Li S, Sun J, Kunwar A, Miao W, Hao J, Bao Y (2014) The study of interficial reaction during rapidly solidified lead-free solder Sn3.5Ag0.7Cu/Cu laser soldering. In: 2014 15th international conference on electronic packaging technology
41. Yang M, Yang S, Ji H, Ko Y-H, Lee C-W, Wu J, Li M (2016) Microstructure evolution, interfacial reaction and mechanical properties of lead-free solder bump prepared by induction heating method. J Mater Process Technol 236(Supplement C):84–92
42. An L, Ma H, Qu L, Wang J, Liu J, Huang M (2013) The effect of laser-soldering parameters on the Sn–Ag–Cu/Cu interfacial reaction. In: 2013 14th international conference on electronic packaging technology
43. Shang PJ, Liu ZQ, Pang XY, Li DX, Shang JK (2009) Growth mechanisms of Cu_3Sn on polycrystalline and single crystalline Cu substrates. Acta Mater 57(16):4697–4706

Printed by Printforce, the Netherlands